DATE DUE

APR 1 4 2014

THE GREAT TEXAS WIND RUSH

NUMBER SIX

Peter T. Flawn Series in Natural Resources

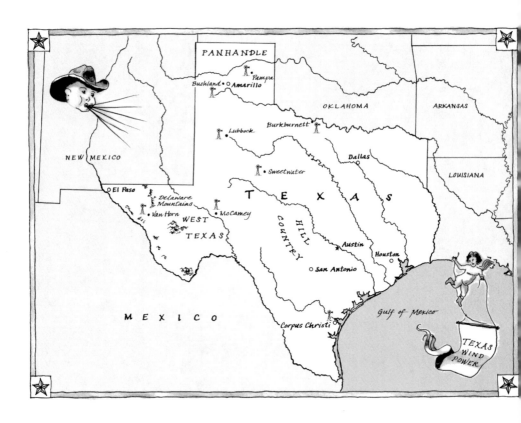

How George Bush, Ann Richards,

and a Bunch of Tinkerers

Helped the Oil and Gas State

Win the Race to Wind Power

THE GREAT TEXAS WIND RUSH

BY KATE GALBRAITH
AND ASHER PRICE

University of Texas Press
AUSTIN

The Peter T. Flawn Series in Natural Resource Management and Conservation is supported by a grant from the National Endowment for the Humanities and by gifts from various individual donors.

Map by Molly O'Halloran, Inc.

Requests for permission to reproduce material from this work should be sent to:
Permissions
University of Texas Press
P.O. Box 7819
Austin, TX 78713-7819
http://utpress.utexas.edu/about/book-permissions

♾ The paper used in this book meets the minimum requirements of ANSI/NISO Z39.48-1992 (R1997) (Permanence of Paper).

Library of Congress Cataloging-in-Publication Data
Galbraith, Kate (Catherine)
The great Texas wind rush : how George Bush, Ann Richards, and a bunch of tinkerers helped the oil and gas state win the race to wind power / by Kate Galbraith and Asher Price. — 1st ed.
   p.   cm. — (Peter T. Flawn series in natural resources ; no. 6)
Includes bibliographical references and index.
ISBN 978-0-292-73583-5 (cloth : alk. paper)
1. Wind power—Government policy—Texas.   2. Renewable energy sources—Texas.   I. Price, Asher.   II. Title.
TJ820.G35   2013
333.9′209764—dc23   2012044363

doi:10.7560/735835

CONTENTS

ACKNOWLEDGMENTS

Wind is intangible, and we needed a lot of help from a lot of people to grasp it. Thanks, foremost, to the dozens of people we interviewed, especially Michael Osborne, Joe James, and Jay Carter Jr. Each was Buddha-like in patience as we asked, again and again, just one more question. Ken Starcher at the Alternative Energy Institute (AEI) and Nolan Clark, now retired from Bushland, acted as a kind of weather vane, pointing us in the right direction. Don Graham at the University of Texas at Austin helpfully suggested pieces of literature to round out our story. Russel Smith let us hold onto his invaluable archives for months, Coy Harris allowed us to return repeatedly to dig through his collection, and Janet Sawin lent us her magnificent dissertation. Was there ever a journalist who had a better grasp of Texas electricity deregulation than Bruce Hight? His incisive work in the 1990s — and the box of old notes he so helpfully pulled from his garage for us — served as a lodestar of good reporting.

Thanks also to the helpful folks at the Briscoe Center archives and the Texas State Library and Archives Commission (especially Tonia Wood).

Thanks, too, to our editors at the *Texas Tribune* and the *Austin American-Statesman*, who supported this effort and, more broadly, allowed us to develop our skills as Texas energy reporters. A boiled-down version of this book appeared in *Texas Monthly* in August 2011, thanks to Jake Silverstein and Jeff Salamon, who imaginatively shaped our work into a readable narrative. Robert Righter generously read this manuscript and offered ideas, as did another reviewer whose name we don't know but who has our gratitude. Our editor at the University of Texas Press, Casey Kittrell, has been a wonderful partner on this project. In so many ways, he made sure the wind was always at our backs. Thanks, too, to Kathy Bork for her close reading.

Asher adds: I would like to thank Kate for inviting me along on the project. She told me she needed someone to make her get up and go on a run each morning. As it turned out, not only did she outrun me, she also made sure our laces were tied correctly. And especially to Rebecca Markovits, who puts me in mind of a Magnetic Fields lyric: I need her "like the wind needs trees to blow in."

Kate adds: Special thanks to Asher for setting the standard on writing—graceful, eloquent, with always a magical analogy or turn of phrase. And thank you to my family for always being there to listen. I'll always remember an outing to California in which my one-year-old niece, too, was entranced by windmills.

INTRODUCTION

More than two hundred million years ago, a great sea stretched over much of the land now known as Texas. That sea, long vanished, has left a lasting imprint on the desolate West Texas terrain. Awful and exhilarating, the land is a smoothed-out, washed-out territory that seems as if it has been ground down by a giant mortar. Apart from mesas here and there or the Big Bend peaks and the Guadalupe Mountains, born of an ocean reef, there is nothing. It is a hard, unromantic land of caliche and scrub, of tumbleweed and bluestem, of flatness and endless sky.

The northwestern reach of Texas, including the Panhandle, is flatter still. Long ago, after the ancient seas retreated, an enormous network of rivers and streams flushed a steady wash of gravel, sand, and mud down from the newly born mountains today known as the Rockies, helping to form the vast expanse of the Great Plains. The sixteenth-century conquistador Francisco Vásquez de Coronado, encountering the southern end of those plains in the vicinity of the modern-day cities of Amarillo and Lubbock, wrote that they offered "not a stone, nor bit of rising ground, nor a tree, nor a shrub, nor anything to go by."

The physics of the Texas terrain—the unobstructed space, the smooth, frictionless earth—make it just right for wind, now as relentless a force on the landscape as the briny waves once were. The state's highway department, in its own tacit acknowledgment of the emptiness its roads cut through, set the speed limit at 80 in these parts, and only then, it seems, because one had to be set. A visitor hurtling along the interstate, as flat and broad as an airport runway, its dark asphalt just short of melting away into the surrounding dirt in hot summer months, feels buffeted by the wind. "Adrift on a billowing ocean of land" is how the late journalist A. C. Greene describes the feeling of bumping along West Texas roads in his memoir *A Personal Country*.

So relentless was this wind that it was said to render women—always women—crazy. In the first pages of *The Wind*, a melodramatic 1925 novel, the naïve, pretty eighteen-year-old Letty Mason ("blond and wavy hair," "eyes as blue as periwinkles," "cheeks delicately pink as the petals of peach blooms"), train-bound to the West Texas hamlet of Sweetwater to be a governess, learns from Wirt Roddy, a mysterious mustachioed passenger, of the hard life that awaits her. "'Folks say the West is good enough for a man or a dog, but no place for a woman or a cat,' Roddy murmured."

> *"But why, why?"*
>
> *"The wind is the worst thing."*
>
> *She drew a relieved sigh. "Oh, wind? That's nothing to be afraid of."*
>
> *He went on as though she had not spoken. "It's ruination to a woman's looks and nerves pretty often. It dries up her skin till it gets brown and tough as leather. It near 'bout puts her eyes out with the sand it blows in 'em all day. It gets on her nerves with its constant blowing—makes her irritable and jumpy."*
>
> *She gave a light, casual gesture with one hand. "It blows everywhere, I reckon, even in Virginia. Sometimes in winter we have regular storms of wind and rain. But we don't think anything of them."*
>
> *He gave her an amused sidelong glance, and*

> *twisted his mustache in silence. His air was that of*
> *an adult who disdains to attempt to make anything*
> *clear to a persistent but silly child.*

Decades after that novel's publication, Gail Caldwell, a writer who hails from the High Plains of the Texas Panhandle, where the winds blow even more relentlessly than those around Sweetwater, remembered being "haunted by the stories of pioneer women driven mad by the wind." This was a force so strong that it would disfigure trees; mesquites were permanently bent over, like women hunchbacked from the day-in day-out strain of drawing water from a well. Cattle could die— drown, really—from inhaling snow that blew horizontally, with tremendous force, during a blizzard. In the 1930s the Panhandle was Dust Bowl territory: the wind was so bad, and the land so dry, that whole fields were scooped up and sent swirling across the nation. Trees were such a rarity that they "grew so singly in yards that a kid might name them," in the words of Texas writer Lisa Sandlin.

Seemingly hard-bitten men, the sort featured in the literature of Texas and the High Plains, learned to fear the wind. After exiting the train station in Amarillo, Gideon Fry and Johnny McCloud, two characters from Larry McMurtry's *Leaving Cheyenne* who had journeyed hundreds of miles to go cowboying in the Panhandle, got a shock. "We got off the train at the big brick station in Amarillo, and it was like getting off at the North Pole," Gid recounts. "The wind whistled down those big streets like the town belonged to it, and the people were just renters it was letting stay." This fierceness also startled a real-life character, Robert E. Lee, stationed in Texas before the Civil War, who bemoaned in a letter to his daughter Mary how the "violence of the wind requires all my hands to hold the paper."

But Texans have long made use of an unrelenting land. More than a century ago they struck out and prospected an unforgiving ground, giving the world the gusher at Spindletop and generations of black gold. And now they have mined the wind that pitilessly blows across the plains. For it was in these lonely, godforsaken patches of earth that the greatest wind-power revolution that the world had ever seen began. In 1981 Michael Osborne, a wavy-haired Panhandle native who gave up a promising career in music promotion to try to save the world, planted the state's first wind farm—the "second-largest wind farm in the known universe," as he described it a few years later to the *Dallas Morning News*—with a handful of turbines on his cousin's ranch in Pampa, a town forgotten to most of the world save for a Dust Bowl ballad by Woody Guthrie.

Even before Osborne began poking around, plenty of wind machines were already scattered around Texas. They were rusty, creaky whirligigs that lingered from an era when ranchers across the Great Plains needed them to pump water so cattle could drink and gardens could grow. The first were installed as the railroads were laid across Texas, to harvest water for the new steam locomotives as they chugged along. You can still find the water windmills in remote pastures crowded with weeds where the steel meets the earth. A Texas company actually remains in the business of manufacturing them, and a few men still make a career out of fixing them through sun and snow and hail.

But the old windmills are now dwarfed by the electric big boys, like the steel GE 1.5 and the Vestas V82, which can each provide enough power for several hundred homes. Some of these new turbines are as tall as football fields are long, and forests of them have sprouted atop bleak, scrubby West Texas mesas. As of 2012, the state had more of these sorts of turbines than all but five countries. The area around Sweetwater, where the fictional Letty Mason would indeed go mad because of the wind, is now home to thousands of them. At night the hills become a canopy of red blinking lights to warn off airplanes. Some of the machines tower over horse-headed pump jacks, feeding landowners two streams of royalties.

The rush to build them has intensified over the last decade and a half, after George W. Bush, then a mere governor burnishing his environmental credentials ahead of a run for higher office, signed a sweeping electricity deregulation law in 1999 that established a statewide requirement for renewable energy. In 2006 Texas blazed past California to become the wind-power leader of the United States, and now it has about two and a half times as much wind capacity in place as the runner-up state, which, strangely enough, is Iowa. While the country as a whole got 3 percent of its electricity from wind turbines in 2011, the Texas electric grid got more than 8 percent of its energy from them—and that figure can soar to 20 percent or more on a spring or fall night, when the winds are especially fierce and electricity demand is low. That Texas could make the improbable leap into wind inspired other states to try their hand at it, too, for if the oil and gas state believed in wind, perhaps there was something to it.

Yet Texans are not blind to the challenges of roping themselves to this new and peculiar source of energy. Wind is fickle, and the state has no cost-effective method, yet, for storing large amounts of electricity to make up for when it slows. When the electric grid needs power the most—in the late summer afternoons when air-conditioners are on full blast—the West Texas breezes tend to come to a near standstill. Winds over the Gulf of Mexico blow more consistently in the heat of the day, and a handful of wind farms have sprung up along the coast, though the biggest prize of all, offshore development, remains elusive because of the high cost of building over water.

There are other problems, too. Even in Texas, there are skirmishes over noise and birds and interference with military radar. The construction of giant transmission lines needed to ferry power hundreds of miles from remote, windy landscapes to big cities has sparked an uproar because they cost billions, and few Texans want ugly wires strung across their land. Wind's economics do not work without a tax credit from the federal government, an issue that gives heartburn to libertarian Texas politicians who favor rural development but abhor handouts from Washington.

On the face of it, then, the oil and gas state leading the nation in wind power appears to be a great energy irony. And, indeed, the oilmen of West Texas gamely battled the windmen in the early days by trying to argue county officials out of granting them key abatements on property taxes.

But the growth of wind in this desolate land actually makes perfect sense, because derricks and pump jacks have already scarred the landscape for generations and whetted landowners' appetites for royalties and tax breaks. "In Texas, they want you to respect their property, but they're used to using it as a commodity," says Patrick Woodson, who got started in Texas wind farming in the 1990s, long before it was seen as a sensible business. Texas ranchers, developers, electricity salesmen, and assorted hangers-on all know good money and a rich resource when they see it, in this corner of the world once described by Georgia O'Keeffe as having "terrible winds and a wonderful emptiness."

Sheer size gives Texas an almost inevitable edge in the wind race. "I thought I knew Texas pretty well, but I had no notion of its size until I campaigned it," Gov. Ann Richards is said to have quipped. Driving end to end across Texas, north-south from Spearman to Corpus Christi or east-west from Beaumont to El Paso, takes at least twelve hours, not including stops. There is plenty of room for wind turbines in the massive West Texas expanse, which already contains the country's largest battery, a radioactive waste dump, a nuclear weapons complex, and all manner of industrial oddities. Many Texans think the turbines are actually an improvement to the dry, scrubby landscape, because they break it up and give drivers along the lonely highways something to look at.

Indeed, it doesn't hurt that Texas is, frankly, less beautiful than, say, California, where towering, snow-capped mountains and gorgeous coastlines have given rise to an environmental activism that's basically lacking in Texas. If a few birds die in the pursuit of energy, not many Texans will care. Jerry Patterson, Texas's colorful land commissioner who pilots his own plane, carries a gun in his boot, and has (so far in vain) promoted offshore wind farms, likes to joke that even if wind farms in the Gulf of Mexico slice up migratory flocks, "after several generations we'll have smarter birds." It's a paradox nicely summed up once by Michael Webber, an energy expert at the University of Texas at Austin: "In Texas, because we don't care about the environment, we're actually able to do good things for the environment."

What this boils down to is that the land in Texas is easy to build on. Because Texas joined the union as a sovereign nation, with most of its holdings already carved up among ranchers and pioneers, more than 90 percent of the state remains privately owned, unlike many of the large, windy states farther west, where the federal government controls large chunks. In those states, energy developers, whether they want to drill for oil or erect turbines, must go through extensive regulatory hurdles before putting up a project. Not so in Texas. "In Texas, you can put anything you want on your own private land, and nobody can say a thing about it," says Randy Sowell, a Texas wind industry veteran raised in Lubbock. Texans look with wonder at states like Massachusetts and the perpetual snarls faced by would-be developers at Cape Wind.

The way these Texans view their land, in other words, differs considerably from the way that Californians or New Yorkers or even Coloradans view theirs. Today's 26 million Texans not only have a large appetite for electric power of all types as the state grows and changes, but they also come branded with a just-do-it attitude, with deep roots in rural self-sufficiency. The state's economy is built on a proud culture of extracting natural resources; the wind is there, so why not harvest it, too? "The wind business doesn't compete against anything else," says Texas landman Chris Crow. Cattle raising, drilling for oil, hunting, and just about anything Texans want to do with their property can happen in between the turbines.

This is a story of how the new energy got started. It is a tale about men and technology, money and politics. It is about how a state flush with oil and gas came to create a source of energy that would be hailed as clean and green. It is about tinkerers and dreamers who spanned a generation, who were short on money and long on philosophers like Buckminster Fuller, and headed to the hinterlands to see if they could figure out a new way to turn on the lights. Environmentalists and Enron executives, Ann Richards and George W. Bush, stodgy utilities and fresh-faced entrepreneurs all make an appearance.

It is a story that reaches back more than a century, even before the oil rush, when West Texas ranchers and railroad men and townspeople began ordering windmills out of mail-order catalogs so they could harvest clean water from the earth. It continues into the 1970s, when space-age engineers like Jay Carter Sr. and his wunderkind son, Jay Carter Jr., began fiddling around with a blade design they had been fashioning for high-performance helicopters. It picks up with Michael Osborne, who hauled a few of those Carter wind machines into the middle of nowhere and plugged them into the electrical grid, and with a Lubbock priest named Father Joe James, who believed that wasting energy was the opposite of godliness. Later came billionaire Sam Wyly, a friend of George W. Bush's, who got into the wind business after his eleven-year-old daughter complained to her daddy about the dirty air. It is about men in the dreariest, windiest parts of Texas who ignored looks from family members who thought they had lost their minds and set about experimenting with the new machinery and fixing it when it broke down—with very little help from anybody. In the end, they would tame the terrifying force that once drove pretty Letty Mason mad and put it to good use. They would, in short, succeed on a scale beyond their wildest imagination.

CHAPTER 1

FOLLOWING A GLIDER

The wind rush that would inspire a nation began small, with a 10-cent glider, the kind that country kids used to buy at the store on days when they wanted a little adventure.

Michael Osborne was that sort of kid. He was sturdily built with a mop of dark hair, and he grew up in a land so remote and desolate, 600 miles from the sea, that it was one of the last parts of the continental United States to be settled by nonnatives. The land was called the Texas Panhandle, and apart from a few gentle hills and cliffs and canyons, it was flat, high, and windy. It was home to vast herds of cattle and fields of wheat and cotton and old oil derricks, but for high school kids, there just wasn't much to do except get drunk, make out, play music, and dream of leaving.

But there were gliders.

One afternoon in the early 1960s, Osborne and a high school pal—Smisson Mulkey Goodlett III, the drummer in his band—bought the basic balsa-wood models and took them to a park in their hometown of Pampa, where the gliders could fly freely on the High Plains. The boys removed the gliders from their plastic wrappings and pieced them together—the fuselage, two sets of wings, and the rudder. When he was ready, Osborne pointed his into the relentless north wind and cast it away.

"It did what all gliders do," he later wrote. "It went down, picked up speed and lift, and then went up and over upside down." It headed down, brushed the top of a wildflower stalk, lifted back up, and then threatened to stall. Then, miraculously, it resumed climbing once more.

It did this a few times, and his friend began to take notice. This was good stuff for a dime-store glider.

"What the glider would do is it was going into the wind and it would stall, and then it would dive, and then it would go up a little higher, and then it would stall, and go up a little higher," Osborne later recalled. "But it always kept its head into the wind. It was the equivalent of throwing a nickel on the floor and it standing on edge, because it was just perfectly balanced and never peeled out left or peeled out right."

The boys hopped in their car and chased the glider to the edge of town as it rose higher and higher. Finally it swooped up and out of sight, up into the vast sky above the plains.

So awed were the two of them that they swore then and there not to tell a soul. And for many years, they didn't.

"It was one of those sort of space moments," Osborne says. "We weren't sure that there weren't space alien beings somewhere. . . . It was unquestionably something that affected me very deeply, in a very fundamental way."

Two decades later, when Michael Osborne set out to build the first wind farm in Texas and the second one in the country, he knew exactly where to put it.

Osborne's family goes nearly as far back as any in the Panhandle, which isn't far in the scheme of things. The state of Texas, which joined the union in 1845, was pretty new itself back then, and for a long time nobody had much idea of what to do with this stump of a place, barren and forbidding, that stuck up awkwardly into modern-day Oklahoma. The Panhandle got its shape from the Compromise of 1850, Henry Clay's desperate effort to slow the nation's march toward war, which handed the Panhandle, El Paso, and $10 million to the fledgling state of Texas, which was still figuring out its boundaries and trying to settle debts from its days as an independent republic. The Panhandle could reach north to the latitude of the Missouri Compromise, lawmakers in Washington agreed, thus ensuring that Texas, a slave state, did not creep north into free country. Texas congressmen also staved off northern efforts to slim down the Panhandle by imposing a more easterly western edge on it. In return for all this, Texans reluctantly gave up their efforts to hold onto other lands that lay to the west along the Rio Grande; these got awarded instead to New Mexico.

For Samuel Claypool Osborne, Michael Osborne's great-grandfather, the Panhandle's very remoteness may have been its biggest draw. Samuel was a teenage soldier in the Confederacy who had been captured and put in an Ohio prisoner-of-war camp. As the family legend goes, he was released even as the Civil War still raged after a cousin wrote Abraham Lincoln for leniency. After the war, times were hard in Georgia, where he'd grown up, so in 1891 he packed up his wife and ten children, some of them named for American presidents and queens of England, and headed for a North Texas town called Weatherford. After another move to Oklahoma, they returned to Texas in 1903, drawn by tales of a wilderness cut through by fast-flowing rivers. They would face no real threat from the Comanches and other tribes, who had been crushed through a series of brutal skirmishes that intensified after the end of the Civil War. And so the Osbornes ended up in the Panhandle, where they would stay.

Samuel Claypool Osborne's move from the Old South to the newly opened frontier lands was a common one. "GONE TO TEXAS was the sign you scrawled and planted outside your house when, like Huck Finn, you were lighting out for the territory, even if you didn't know where you were headed," writes Gail Caldwell, the Panhandle-born author. Her grandfather, a one-armed Civil War veteran, left Tennessee also bound for the Panhandle, probably arriving a little before Osborne's great-grandfather.

What the Panhandle newcomers found were plains covered with short prairie grasses, like blue grama and hairy grama and buffalo grass, as well as vine mesquite as far as the eye could see, bounded to the east by a caprock escarpment. This was the southern edge of the Great Plains, the ocean of flatness that cleaved America. The new land was stricken with what Caldwell calls "Old Testament weather." The sun could give way to terrifying hail in an instant, and twisters swept across the High Plains from any direction with nothing to stop them. The first settlers in the Panhandle lived in dugouts, homes with dirt floors literally dug out of the ground, where they faced spiders and snakes but had protection from the cold winters and constant winds. Gray County, where Michael Osborne grew up, was especially prone to fearsome winds due to its perch on the edge of an escarpment that drops hundreds of feet. Pampa itself sits at an elevation of about 3,200 feet—a midpoint in the Texas High Plains, which tilt imperceptibly from 4,700 feet in the northwest down to 2,000 feet in the southeast, over the space of a few hundred miles.

Samuel Claypool was undeterred by the hail and the blizzards. Having found land beside the Canadian River, he settled down to ranch and preach until, a few years later, drawn by the comforts of town, he sold the ranch and opened a dry-goods store in the small railroad town of Miami, Texas, northeast of Pampa. His youngest son, Johnson Polk Osborne, joined him in the business, and together they sold clothing and coffins and plenty of things in between.

One of Samuel Claypool's grandsons, Michael Osborne's father, began one of the first big feedlot operations in the Panhandle; you could find your way to the place from miles off just by smelling. Another of Samuel Claypool's grandsons, Jake Osborne, who was Michael Osborne's uncle, ran the Pampa Lumber Company. In the early days, after oil was discovered in the Panhandle in the 1920s, it sold wood, which was needed for houses, oil derricks, and railroad ties. Between 1926 and 1931 Pampa's population jumped from a mere 987 to over 10,000, and the town earned standing as the seat of Gray County.

When Michael Osborne was growing up, the Pampa Lumber Company also sold something else, something that helped farmers harness the winds of the Texas Panhandle and put them to work growing vegetables and fattening cattle. It was a tall steel whirligig called the Aermotor.

The winds of the High Plains, sweeping down relentlessly from the north one day and from the south the next, shape the land, and the character of those who live there, as much as the soil and the grasses and the flatness. Georgia O'Keeffe, who spent several bewildering but happy years around the time of the First World War in Amarillo teaching drawing to high school and college students, recalled later that the winds were so intense that "sometimes, when I came back from walking, I would be the color of the road."

The winds that could blow shrieking storms and drive huge prairie fires also sparked wild, fantastical ideas. Pioneers, tiring of the slog across endless flat en route to California, began conceiving of "wind wagons"—coaches with sails instead of oxen, racing east to west, and then back, across the plains. A few set off, but alas, the winds against the sails could not haul wagons out of gullies, as they lifted ships from troughs in the sea, nor could the wagons continue their journey when the wind was calm, or in the wrong direction. So the horses and oxen stayed on, for a journey that was slower and less of a thrill, but steadier.

During the awful Dust Bowl days of the 1930s, the winds robbed the people of the land. They picked the soil up and carried it in a big dark cloud that rolled across the plains. The arid earth, the sod cut for decades by farmers who nearly broke their backs tilling it, flew through the sky and choked it with black, so that people made footprints in the dirt even in the safety of their own houses. "Not much a housewife can do while the storm lasts, only to walk from room to room, muttering, 'Why did I ever come to West Texas?'" wrote Nellie Witt Spikes, a Panhandle farm woman born in 1888 who stayed through storms that literally lifted up her chickens' wings and carried the animals away, even as the rains failed to come. The black clouds swept from west to east, depositing soil as far away as Washington, D.C., as Timothy Egan recounts in his memorable tale of this era, *The Worst Hard Time*. Roughly a third of the Panhandle farmers left during the Dust Bowl, abandoning their tough life for the promise of places like California. The rest, too poor, too weak, or too attached to leave, found solace in each other.

They also found solace in the music. Woody Guthrie, age seventeen, had wandered into Pampa in 1929 with little more than a harmonica and a suitcase. The town was "wilder than a woodchuck," its streets lined with makeshift shacks built by oilmen who made piles of money and then spent it on whiskey and slot machines. He stayed for eight years, through the worst of the Depression and the drought and the dust, cycling through jobs hawking moonshine in a drugstore, painting signs on shop windows, and playing the guitar at square dances for a few dollars a night. He was still in Pampa, with a wife and a baby on the way, when the worst dust storm of them all rolled through one April day in 1935. It was thousands of feet high, a towering black wall, and it arrived, Guthrie later recalled, "like the Red Sea closing in on the

Israel children." Everybody fled for shelter, and "it got so dark that you couldn't see your hand before your face." An electric light bulb in that dust-smothered room offered such little illumination that it resembled the glowing end of a cigarette.

Written years later, the ballad went thus:

> *This dusty old dust is a-gettin' my home,*
> *And I got to be driftin' along.*
> *A dust storm hit, an' it hit like thunder;*
> *It dusted us over, an' it covered us under;*
> *Blocked out the traffic an' blocked out the sun,*
> *Straight for home all the people did run . . .*
> *So long, it's been good to know yuh. . . .*

Osborne knew an extra verse from his boyhood, one that speaks of the town where Guthrie cowered and prayed in a dark room on April 14, 1935, the day known to history as Black Sunday:

> *It was in the town of Pampa*
> *The county of Gray*
> *Nothing was standing,*
> *It was all blown away.*
> *So long, it's been good to know ya."*

Michael Osborne had not been born when Black Sunday struck, but big dark clouds of land still sometimes barreled through during the 1950s, when he was a young boy. "That was one of the memories I had as a child," Osborne says. "I was walking home from school, and I looked on the horizon, to the north, and I saw this line, this brown line, and I knew that I needed to get home in a hurry because a great dust storm was coming. And I ran like crazy—I was probably in the third grade— ran like crazy, and made it just in time before the sixty-mile-an-hour wind and the dust came. And it was just like a blizzard, except it was dust. There would be sand dunes to the eaves as if there had been a giant snow," he says.

The Aermotor windmills sold by Michael Osborne's uncle stood tall through the dust and the dirt. And they could take the winds, the every-day winds that swept unobstructed across the High Plains, and turn them to a useful cause.

The concept of the windmill is at least a millennium old, and perhaps much more. The first known machines, for grinding grain and drawing water, appeared in tenth-century Persia, and the concept quickly spread to other nations. Those early wheels spun on a vertical axis and resembled, in the description of wind historian Peter Asmus, a sort of "merry-go-round." Middle Eastern civilizations continued to refine them, and in the twelfth century returning crusaders may have introduced the concept to Europe. By the seventeenth century, the Dutch, whose lowlands lay precariously close to the sea, had erected thousands of windmills to pump out water as part of massive land reclamation projects. By then the machines had evolved from merry-go-rounds to heavy, house-like contraptions with sails, which spun around on a horizontal axis, like the large modern turbines of today; clever craftsmen also added fantails, allowing the machine to rotate to make maximum use of the breeze. Aside from what farm animals or human muscles or burning wood could do, the windmills were the key mechanical source of energy for Europe, where as many as 100,000 existed near the turn of the twentieth century—just as advanced steam engines, fueled by vast supplies of newly tapped coal, ushered in the Industrial Revolution.

As prominent pieces of architecture on a flat landscape, these windmills were used for other purposes as well: the blades, when stopped in a certain position, could alert passersby to important events in the farmer's life, such as a wedding or a new baby. When World War II arrived, anti-Nazi fighters in the Netherlands broadcast coded messages via windmill.

In the Texas Panhandle in the late-nineteenth and early-twentieth centuries, the purpose of the early windmills was singular: to pull up water from aquifers far beneath dry soil. Only eighteen inches of rain fell annually on average at the Panhandle's western edge, so underground water, pumped from tens or hundreds of feet down, was needed to tame the land. The house-like Dutch windmills, which ground grain, were too costly, large, and labor-intensive for mass-market water pumping in the Great Plains. So in 1854 a Vermont-born mechanic named Daniel Halladay, who had once served as an equipment manager at the federal armory at Harper's Ferry, invented the American farm windmill, designed to pull up water more cheaply and easily than its forebears.

"Now one of the greatest motive powers of nature is utilized, and, as it were, made subservient to the will of man, conferring untold benefits to millions," a nineteenth-century Midwestern account explains. Ease of shipping was crucial, too, and soon Halladay was sending windmills across the plains. Business got serious after the Civil War, and by the 1870s factories in Illinois, which were closer to their market than was Halladay's home base in the East, were busy churning out his windmill and competing models en masse.

The technology was deceptively simple. A stiff breeze turned their blades—Halladay's had flat wooden ones—which were mounted high atop a tower. As the wheel spun, it drove a gearbox that produced an up-and-down motion that allowed for pumping and ultimately pulled water from deep under the earth. A vane, or tail, swung around with the wind to keep the machine pointed in the right direction.

In Texas, which by its sheer size and rural makeup would in time become the nation's largest market for windmills, the first industry to adopt windmills was the railroads, which needed clean water to generate steam. Windmills for locomotives were erected in Texas starting in 1860, on a sixty-mile stretch running southwest from Houston, though this line was abandoned after the onset of the Civil War. By the early 1880s the railroads stretched into West Texas, aided by powerful windmills called Eclipses, whose wheels could reach eighteen feet in diameter. The Eclipses, paired with large holding tanks for water, began appearing every thirty miles or so along the tracks. A national sales agent for the Eclipse opened an office in Dallas in 1908, and it "sold more windmills than most of the company's offices combined," according to geographer Terry Jordan.

The cities needed windmills, too, to provide a water supply for the people who were drifting west with the railroads. Often, then, the towns of West Texas appeared "as a forest of wooden towers and whirling wheels" rising up from the desolate plains, in Jordan's description. Midland, these days better known as the oil and gas capital of West Texas, where future governor and president George W. Bush got his start, was known in the early 1900s as the "Windmill Town," because, according to the *Handbook of Texas Online*, "virtually every house had a windmill in its yard."

The windmills could not pump enough water to support acres of thirsty crops, but they did bring up enough to keep cattle, so ranches small and large formed in Central and West Texas, on lands that previously had been too dry to attract much commercial activity. "I regard the invention [as] a national blessing," a man named W. P. Gillespie from the cotton town of Tehuacana, east of Fort Worth, wrote to a windmill company in 1875. "Doubtless it will be the prime agent in rendering the wide and desolate prairies of our western border habitable, and aid men in reclaiming and bringing into cultivation the arid sands of the desert."

In a few places out west, the windmills were later even put to work retrieving another substance from deep in the earth. T. Lindsay Baker, probably the country's foremost windmill historian, reports that "the most common liquid other than water that windmills pumped was petroleum." His book *American Windmills* includes a photograph of an Eclipse windmill serving as an oil pumper near the West Texas town of Toyah in 1911.

Windmill agents rode from place to place hawking their wares, and by the 1890s, with western growth seemingly boundless, Sears Roebuck and Montgomery Ward had added windmills to catalogs that were already thick with items like watches and hoop skirts and kitchen essentials for American homesteaders. The wheels might cost $20 or $30, not including the tower. When the windmill was delivered, the farmer would often summon his neighbors to help put it together and dig a well deep enough to tap the aquifer, which, hopefully, would not be more than a few dozen feet down. Sometimes a work team would start in with shovels, but when possible they would rope in horses, which ground a drilling bit deeper as they walked in a circle.

On the desolate plains, the windmill had become a beacon of civilization. "Within a short time after its introduction the windmill became the unmistakable and universal sign of human habitation throughout the Great Plains area," historian Walter Prescott Webb writes in his magisterial book *The Great Plains*. "It was the acre or two of ground irrigated by the windmill that enabled the homesteader to hold on when all others had to leave. . . . It made the difference between starvation and livelihood. These primitive windmills, crudely made of broken machinery, scrap iron, and bits of wood, were to the drought-stricken people like floating spars to the survivors of a wrecked ship."

The mighty XIT Ranch along the western edge of the Panhandle, whose owners had acquired three million acres of land from the state of Texas in exchange for bankrolling the construction of the domed capitol building in Austin that stands to this day, ordered several hundred

1.1. *A water-pumping windmill being erected in the Texas Panhandle c. 1900. Claude Walter Crowell, grandfather of the present-day windmiller Mike Crowell, holds the horse. Courtesy Mike Crowell.*

windmills, which rose to an average height of 34 feet and pumped wells as deep as 400 feet. The ranch—the "largest range in the world under fence," bigger than some Yankee states—also boasted, reputedly, the tallest windmill in the world, a wooden contraption that rose 132 feet in the air, the height of a very tall pecan tree, so that it could catch the breeze above the walls of the canyon it occupied. It blew down in 1926.

J. B. Buchanan, a Panhandle farmer born in 1906 who spent much of his life collecting antique windmills, recalled driving with his parents in a Model T ninety miles from Spearman, near the Oklahoma border, to Amarillo: "We'd drive through everybody's farm and at every gate was an Eclipse, 102 of them, for watering livestock. And at every one of them we'd stop to water ourselves and our Ford. How good that water tasted on a hot dusty day!" In her novel *That Old Ace in the Hole*, Annie Proulx gets at the dependability of the windmills on the plains: The ranch hand Tater Crouch tells the Dutch windmiller Habakuk that he "was not impressed with the slender stream of water that issued from the pipe of a well they had sweated on for days.

> *'Hell, I can piss fastern that thing can throw out.'*
> *'But you can't do it for so long,' said Habakuk*
> *and knew [the ranch hand] would not make a*
> *windmill man."*

Texas was said to be the "best market area for windmills in the United States," according to Jordan. The windmills were especially crucial after barbed-wire fences, one of the country's major post–Civil War innovations, began carving up the plains and cutting ranchers off from the streams and springs. Often they were the only vertical structure on the plains for miles around; according to T. Lindsay Baker, ranchers' wives sometimes climbed them and put kerosene lamps at the top, like landlocked lighthouses, to guide their husbands home on dark nights.

The oldest windmills, from Halladay's era, were made of wood, but the technology rapidly changed and improved. An engineer named Thomas O. Perry made the improvement of the windmill his obsession, running more than 5,000 tests before he came up with one that contained metal blades curved to resemble a pinwheel. Perry had been working for U.S. Engine & Pump Company, Halladay's old shop, but executives there were not interested in the new design. So he took his ideas elsewhere, and in 1888 the Aermotor windmill was born in Chicago. This new creation, which competitors belittled as the "mathematical windmill," had extra pumping power and was a runaway success. By 1892 the farmers of the Great Plains had bought 20,000 Aermotors, which became the leading brand in Texas, despite the presence in the state of a scattering of manufacturers like the F. W. Axtell company in Fort Worth. A big windmill distributor, Burdick & Burdick, opened its doors in El Paso in 1927.

"My family had been in the wind business, in the Aermotor business, for generations," Michael Osborne says. "And if anything settled the Texas Panhandle, it was the invention of the Aermotor, which was the old Chicago Aermotor. It was a reliable windmill that could pump water up, because we had a good water table there, and if you got the water up, and got your herds watered, then there was grass there and you could actually raise cattle." Eventually Michael Osborne would realize that the wind machines sold by his uncle to thirsty farmers could, by using the same type of energy, be used to create something that the earliest Panhandle settlers could barely conceive was possible: electricity.

Growing up in one of the most remote places on Earth, with a daddy who raised cows and an uncle with a lumber shop, just about guaranteed that Michael Osborne would be a tinkerer. But it also guaranteed that he would itch to do something that would take him away.

From about the time he could operate a plow he had worked in his father's land-and-cattle operation, home to about fifteen hundred head. "The worst job at a feedlot is—well, there's a lot of bad jobs at a feedlot—but one of the worst jobs is when you're working cattle, there's a little narrowing of the trough, and then they go into the squeezer chute and stuff," Osborne says about his ranching days. "And so a guy like me, a little guy, a ten-year-old, would be the one that would run the cattle into the working corral, and then you would push them into the wooden chute, and then you'd put bars between each one of them so they're all ready to go in, and then you would punch them up and push them into the chute, where they'd get squeezed and they'd do the dehorning and that sort of stuff. And so, that means you're right down there in with them."

He did not yearn to be a rancher; instead, like Woody Guthrie a few decades before, he craved music. Music was as intertwined with Panhandle life as the winds and the sky. It was a way of filling up the desolate spaces. Guthrie had made the place famous, but by the time Osborne was a kid it was the land of Buddy Holly, where "rockabilly" beats filled the wide-open spaces and cheered the people who drove for miles to hear it.

Holly was born in 1936 in Lubbock, nearly 200 miles south of Osborne's hometown of Pampa, which in that part of the world practically qualified as next door. So desperate were they to make music before an audience in the isolated Panhandle that Holly and his school pals would "play for the opening of a pack of cigarettes," one band member recalled. A few decades after Holly, another Lubbock group—Joe Ely, Jimmie Dale Gilmore, and Butch Hancock—called their country band "The Flatlanders," after the unforgiving landscape of the plains. Joe Ely even wrote a song called "Because of the Wind": "Do you know why the trees bend at the West Texas border? / Do you know why they bend / Sway and twine? / The trees bend because of the wind / Across that lonesome border / The trees because of the wind / Almost all the time." His band mate Butch Hancock tells the joke that in these flat parts of Texas, "you can see for fifty miles, but if you stand on a tuna can you can see for another fifty."

Osborne had been making music of his own. He had saved up enough money to buy a guitar, and in high school he started a band called "The Essex." To him, the name sounded like a place in England—which, of course, sounded cool—plus, "it actually had 'sex' in it." The Essex played rock and roll, which meant that "we had a few songs that we wrote, but we mostly copied the Beatles, the Rolling Stones, and to my dismay Paul Revere and the Raiders," Osborne later wrote. "We were pretty good for youngsters though."

Osborne was no Buddy Holly and knew it, so he decided he wanted to become either an astronaut or a filmmaker. In the late 1960s he made his choice, turning down film school at the University of Southern California to study aerospace engineering at the University of Texas at Austin. At least that was the plan.

In the early 1970s Austin was still little more than a groovy cow town, save the looming pink granite dome of the capitol. Some of the bars on the boozing strip of Sixth Street were still known as saloons, and the *Cactus*, the university yearbook, continued to award the Bluebonnet Belles to the most fetching co-eds—evening wear and elbow-length white gloves in the photo tableaux. It had schools and statues honoring Robert E. Lee, and some of the black neighborhoods were left over from the postwar freedman's villages. That's post–Civil War. The real estate boom and bust was a decade or two away, and Austin's slacker-tude had been jolted only by the 1966 sharpshooting murders by Charles Whitman from way atop the university tower.

"Austin in the early seventies was mellow to the third power, a curious amalgamation of students, hippies, ne'er-do-wells, and politicos—characteristics that might frequently be found in the same individual—and if any one location was the symbolic headquarters for the Austin attitude, it was an old national guard armory turned live music cavern called the Armadillo World Headquarters," writes Rick Koster in *Texas Music*. It was at the Headquarters, opened in 1970, where Willie Nelson drank and smoked his way through the 1970s, where you could catch a Frank Zappa show or Asleep at the Wheel, where a young jeans wearer from Jersey named Springsteen popped up, and where Captain Beefheart, whoever the hell he was, made a name for himself. "In the vision of Texas we want to communicate, the nine-banded Armadillo is more important than John Connally or the oil industry," a 1976 manifesto about the Armadillo announces with the usual bluster. "It's a vision the rest of the country hasn't seen yet."

Osborne had this vision, too, and in his sixth year of college, just three months shy of graduating, he dropped out and was living the dream. He started Directions Company, a publicity agency, and he soon found himself with two of the best-known clients in town—the counterculture stalwart Armadillo and the establishment UT Co-op, which was the university's bookstore and merchandise place on the "Drag." He had gotten the job at Armadillo after schmoozing with Eddie Wilson, the music hall's founder, and bonding over the name Osborne, Wilson's middle name. Your name was good for something then, and soon Osborne found himself booking and directing a stable of artists to create loopy poster art—lots of tie-dye and electric colors and guitars coming out of musicians' skulls. If you needed the honky-tonk's media man, you called Osborne.

He fit right in at the Armadillo, which had a "booking policy [that] reflected the beatific philosophy of its management (which viewed the Armadillo as less a business venture than a giant playpen for stoners)," writes Koster. "Blues, rock, country, *conjunto*, folk bands, and musicians of every description played there, and the amazing and utopian result was that all manners of formerly antagonistic subsets—bikers, hippies, rednecks, acid heads—found themselves dancing, drinking, getting high, and laughing together." When he wasn't working as the media man at the Armadillo, producing radio spots, commissioning posters, selling albums, and trying to persuade fans not to race actual armadillos, he was building a client base around Austin, promoting the bong shops Magic Mushroom and Oat Willie's.

Osborne had the job most Austinites would die for. But he started to fear that he might not make it, literally, if he stayed in the business. Janis Joplin had gotten her start in Austin, and in 1970 she had been found on the floor of her Los Angeles hotel room, dead of a heroin overdose at age twenty-seven. Jimi Hendrix and Jim Morrison were gone, too, and the problems had not spared Austin. "You know, I'd probably already seen at least one drug death by then, maybe two," Osborne says. The music business was also becoming less free-spirited and more businesslike, even in chilled-out Austin, turning Osborne off from the whole enterprise. He had spent a half-dozen years in the Austin music scene,

1.2. *Michael Osborne, seen here in 2011, set up the first wind farm in Texas, in his hometown of Pampa. He still works on renewable-energy projects in Austin. Photo by David Bowman.*

having left UT his senior year to be a full-time adman, but now he found that "some of the soul began to leak out a little bit."

He was also reading a book called *Utopia or Oblivion: The Prospects for Humanity* by the futurist and inventor R. Buckminster Fuller. Fuller favored terms like "Spaceship Earth" and "ephemeralization," and his book held that oil and gas worked as a sort of backup bank account, and humankind needed to learn to live on income energy, like solar. And so Osborne decided he wanted to change the world, which he admits sounds pretty hokey, even for a hippie from the Panhandle who is fond of Don Julio's tequila in round bottles and writing about concepts that Buckminster Fuller would probably have appreciated, like "oneness" and "planet collective consciousness" and our place in the galaxy.

Osborne being Osborne, he chose a circuitous, whirlwind path toward changing the world. In 1976 he started out with solar power, building homes that maximized the use of the sun's energy for cooling and heating without the need for actual panels. In 1980 he opened a business called Osborne Solar, based out of a solar-powered Texaco gas station on the main street of Elgin, thirty miles east of Austin, back before that town had a stoplight. It had a giant solar device on top of the roof that heated water for a car wash next door, and the shop sold the likes of solar flashlights and solar pianos rather than normal service-station goods like potato chips. He had also spent a roundabout few years doing play-by-play basketball for the Taylor High Ducks and leading the Texas charge of the guerilla TV movement, the 1970s' counterculture forerunner to YouTube.

A few years after experiencing a far-out epiphany while watching a comet streak feebly across the West Texas sky, he decided to get real and look for the next big thing, the thing that would drive the twentieth century into the twenty-first. He never forgot what he had learned at his uncle's shop in Pampa, and from the courses in aerospace engineering and business he'd taken before dropping out of the University of Texas. And, of course, he remembered the glider and how it had swooped and dipped and never returned. "Being of an engineering bent," he said decades later, "it was clear to me that wind would be the first renewable technology that would mature and be cost effective." And so, in 1981, thirty-two-year-old Michael Osborne drove back to the dusty town of Pampa.

The wind turbines going up in West Texas today stand as tall as twenty-story buildings, higher than anything else around, with rotors so long—more than 300 feet—that the flatbed trucks that ship the blades and towers through the countryside need police escorts. Each of the biggest machines costs a few million dollars and can produce enough power to keep the lights on in hundreds of homes.

Compared to these giants, Osborne's project was almost laughable. He planted five turbines, like so many flags in the earth, on a ranch belonging to his cousin, a county judge; two others went on adjoining property that belonged to a Coca-Cola distributor named Bob Mack. The machines had two blades each—today's have three—and Osborne bought the machines for a total of about $80,000, minus a $20,000 tax credit from the federal government, from the "pretty boys" at Carter

Wind Systems, a father-son team that assembled windmills in a warehouse near the Oklahoma border in Burkburnett, Texas—when they were not building helicopters. The turbines stood sixty feet tall, which is taller than some lighthouses, with rotors thirty-two feet in diameter, each capable of producing twenty-five kilowatts of energy. It was a tiny amount—enough to power twenty or thirty households, if the wind blew all the time and the machinery functioned perfectly, which of course it didn't.

They spent a week putting them up during the summer of 1981. Because he knew something about camera work, having taken a brief course on holography years earlier and nearly gone to film school for college, Osborne had the installation videotaped. The grainy footage shows the men's long hair—it was the '80s, after all—waving in the stiff breeze, right along with the tall grass. They wore jeans and boots and looked very Dukes of Hazzard, squinting into the cameras. "I'd just as soon start making money," a young Osborne says to the cameras, as men work on the turbines atop the waving grass. "I'm really sick and tired of spending it." Mack, the owner of the other two turbines, just liked watching them: "You look up there, and why it looks like—well it does, it looks like a well-figured woman, the way it's designed and everything, you know," he says in the film. "Makes you look twice."

Then they turned on the turbines. It was, Osborne says, the first wind farm in Texas and the second one in the United States. He had missed being first by only a year: twenty wind machines had just gone up on Crotched Mountain, a ridge in southern New Hampshire, at the end of 1980. Those windmills were supposed to feed power to a home for disabled children, but some blades soon broke off. "Within a year, all of the turbines had been destroyed and the site bulldozed over," writes Peter Asmus in his industry classic, *Reaping the Wind*.

Osborne's lasted longer, but the big problem he faced was lightning strikes, which can crack the blades or, at worst, cause a turbine not properly grounded to explode in a fireball, even (though extremely rarely) today. The machines were essentially tall hunks of metal above the flat Panhandle prairie—and "as any golfer knows, sticking a pole up in the air is a really good way to get some attention from a thunderstorm," Osborne says. He got used to getting calls from his cousin, Gray County Judge Carl Kennedy, after storms rolled through Pampa. At least once he drove out to the ranch with his two assistants, one of whom was a musician from the band Greasy Wheels, only to find what "looked like Star Wars bullet holes," half-inch-wide holes burned in the control box

1.3. *Putting up the Pampa wind farm, which began operating in 1981. Photo by Michael Osborne.*

where the lightning had come through. "It blew everything out of the box," he recalls.

Four years into his wind farm, with the wear-and-tear on the control boxes and never-ending change-outs, Osborne decided to end the Pampa project. He was getting paid about 2.7 cents per kilowatt-hour for the electricity, but "I really needed to be making around a nickel," he says, adding: "I think, ultimately, if we broke even, we just barely broke even." He was guaranteed at least a modest price for the power, thanks to a brand-new federal requirement that electric utilities buy electricity at reasonable rates from small independent producers. That policy, the Public Utility Regulatory Policies Act (PURPA), would prove crucial to the industry's rise—"like populism with electrons," says Osborne, who sold his power to a conservative coal-burning utility called Southwestern Public Service. Even so, the Pampa wind farm was struggling. "We were four or five years into [the Pampa project] at that time, we had put a new transmission in one of them, and the other ones were getting banged," says Osborne. The numbers to upgrade the equipment, he says, "just didn't work out."

Nearly thirty years after Osborne, T. Boone Pickens, the legendary oilman-turned–corporate raider–turned-windman, would also select Pampa as the site for what he grandly announced would become the world's largest wind farm. Pampa reveled in the spotlight: the town where once Woody Guthrie's guitar strings "roar[ed] around in the wind" was by then a place of empty parking lots and half-boarded-up malls, Mexican restaurants and McDonald's, with shops selling "gently worn clothes." Pickens had a ranch nearby, and he, like Osborne, knew what the winds could do. Pampa would become, Pickens trumpeted, the wind capital of the world. He dreamed of putting up thousands of turbines—not on his own land, to be sure—and powering 1.3 million homes as they spun.

It didn't quite work out that way: Pickens got tripped up by various factors, including the credit crisis, falling energy prices, a lack of transmission lines to move the power, and, arguably, his own hubris. He gave up the last of his Pampa wind leases in 2010.

But by then, the Texas windmen were no longer pioneers. They had, with Osborne's help, crafted a way to bring renewable power to Texas that would take it far beyond the fragile speck of a wind farm that had gone up in the fields of Pampa in 1981.

CHAPTER 2

THE TINKERERS

Michael Osborne may have put up the first wind farm in Texas—the first project to plug a set of turbines into the electric grid and earn some money for the power they created—but he was far from the first Texas tinkerer to put up a wind machine that turned on lights. Tinkering, and the self-sufficiency that it implies, runs thick in the blood of Texans. Some of this has the flavor of myth, especially in the modern era of "symbolic frontiersmanship," Larry McMurtry's term from *In a Narrow Grave* to describe how cowboy hats and boots have largely replaced the real deal.

Still, the state's identity rests on the fact that it essentially built itself from scratch, led by men who yanked it—stole it, really—from the Indians and then from the Mexicans and submitted it, more or less fully formed, to the union. It is also an enormous state—the largest in the union until Alaska and its vast wilderness joined—and for well into the twentieth century, Texas was predominantly rural. In 1900 just 17 percent of the state's population lived in the big cities, according to the U.S.

Census, and most of the rest made their livelihood from the land, which in West Texas meant cotton and wheat farms and cattle ranches. What they did, they had to do on their own, because they lived miles from the nearest neighbor, too far to summon help easily.

At the turn of the twentieth century, the tinkering was focused on water windmills. These were arguably a farmer's most precious possession, for the tall towers that drew up water made life possible on those dry, dusty plains. But they were also his bane, for they broke often—no surprise, perhaps, given that they ran day and night, buffeted by wind gusts and storms and attacked by insects, freezing temperatures, blistering heat, and anything else nature hurled at them. Proulx writes in *That Old Ace in the Hole* of

> *leaky tanks, worn leathers, broken sucker rods (in many the copper rivets had been replaced with bent nails), broken furl winch wires, missing sails, helmets ventilated with bullet holes, motor cases in desperate need of oiling, channels plugged with sludge, tailbone pivot bolts worn, washers, rings and bearings worn, bearing bars broken, tail chains broken and wedged in the mast pipe, crows' nests on several non-working mills, stopped because the wheel's vibration had shaken some of the birds' treasures—marbles, bolts, pieces of bone, shiny pebbles—loose from the nest to lodge in the water column and ruin the cylinder.*

Small farmers generally did the work of fixing windmills and the pipelines that ran water to the stock tanks. On the Great Plains, noted Walter Prescott Webb, it was said that "no woman should live in this country who cannot climb a windmill tower or shoot a gun." But at huge cattle ranches like the XIT in the Panhandle, where it could take most of a day just to cross the place, crews called "range riders" or "windmillers" spent their days working out of a chuck wagon, in warm sun or blowing snow, to keep the hundreds of windmills in good working order.

Tinkering with windmills was dangerous work. The Proulx character Bob Dollar dryly observes that they "looked like a combination of tripod and meat grinder," and indeed history is full of deaths and broken necks from all manner of accidents, some of them involving children who tumbled off towers or took ill-advised joyrides on the blades.

In a spirited 1990 essay in *Texas Monthly* called "The Windmill," Anne Dingus recounts a J. B. Buchanan tale of one cowboy who thought that the windmill he had climbed was toppling over (it wasn't) and leapt off into a stock tank. Nellie Witt Spikes, a twentieth-century Panhandle farm woman and writer, tells of the death of one windmill novice who climbed a tower and got his legs paralyzed when the thing unexpectedly began twirling. He spent the night alone in the bitter cold and later died. Mike Crowell, a third-generation windmiller who as of 2012 was still on the job in the Panhandle town of Claude, was working in Palo Duro Canyon one day when his assistant accidentally dropped a hammer on his head from atop a windmill tower. Crowell took off his hat, felt blood, and headed immediately to the hospital, where he would get five stitches. "That was probably the fastest that old pickup come up out of that canyon," he recalls. Crowell's father, during his windmilling days, nearly got his fingers cut off once trying to work on a tower whose wheel didn't have its brake on.

At the XIT Ranch, plenty of cowboys were afraid of climbing the windmills, to the point that, according to Paul Gipe's recounting of stories by the wind expert J. B. Buchanan, "one scared cowpoke exchanged his turn for doing another's laundry." An XIT Ranch foreman was said to test a new worker's loyalty by whether he would climb the windmill to grease it without complaining, T. Lindsay Baker writes. As another tale went, "it was no uncommon occurrence for a cowboy to find it convenient to quit just before his turn came to climb the swaying tower."

A big technological breakthrough came shortly before World War I, when Aermotor and other companies introduced windmills that oiled themselves, drawing from a bath-type reservoir in the windmill. For farmers, "it meant that you only had to check it once a year, rather than having to go up and grease it once a month," says Nolan Clark, a wind specialist in the Texas Panhandle who grew up with a windmill by his childhood home in Cook County, north of Dallas. Today's windmills, remarkably, still have the same basic design that Aermotor has used for more than a century, a simple wheel packed with blades. And indeed, Aermotor is still in existence, though no longer in Chicago: it hammers out a few thousand steel machines per year in a small factory in San Angelo, Texas, which resounds with the clang of metal. "Everything's made in America. We don't even use foreign bolts," says Bob Bracher, a straight shooter running the place more than 120 years after the company began.

But an even bigger breakthrough was at hand. It would be a different kind of wind machine altogether—one that would take the idea of harnessing the winds an extra step and convert their energy into electricity rather than the purely mechanical tasks of pumping water or grinding grain. And it, too, would tap into the ingenuity of isolated Texas ranchers, whose habit of producing what they needed on their own would prove hard to break.

Wind has a lot of kinetic energy. What it does is move—and the way it moves has everything to do with the relationship between the Earth and the Sun. Put simply, the Sun heats up the Earth unevenly, leading to differences in temperature and pressure in the air above the planet, and these differences determine the patterns of air currents. This happens when, for example, land absorbs solar heat and releases it into the air above it faster than the ocean does. Also, parts of the Earth—notably, the tropics—get more sunlight than other parts because of the tilt of the planet as it rotates and swings across its orbit of the Sun.

Meteorologists like to say warm air is buoyant. As air rises, though, it cools down, rushing back down (falling, really) to take the place of hotter air that's rising. Because cold air is denser than hot air, it weighs more and presses down. Put another way, wind is the consequence of natural forces evening out pressure differences in all directions. It's why, at a simpler level, opening the windows in a hot, stuffy house can create a draft that carries air indoors.

In Texas, the winds come from different directions, depending largely on the place and the season. Winds along the Gulf Coast are sure and steady. They arrive from the south—from the sea—because as warm air rises off the land, cool sea air blows in to replace it. And because Gulf Coast summers get oppressively hot, the winds are strongest in the afternoon and the summertime.

In West Texas and the High Plains, the winds are mostly out of the southwest, as cool air from mountain ranges in areas like Far West Texas moves in to replace the hot air that sits over the flatlands. But powerful gales often blow in from the north or northwest, especially during fall, winter, and spring. These winds are more variable than the Gulf winds, because as the air rushes over land, it encounters friction in the form of hills and valleys, and this friction—such a contrast to the smooth sea—causes it to dance and swirl. It bends trees and flings sand. This is no sentimental wind. When the weather goes "all sideways" it can swallow a preacher's words at a San Angelo funeral, as Cormac McCarthy describes in *All the Pretty Horses*. "When it was over and the mourners rose to go the canvas chairs they'd been sitting on raced away tumbling among the tombstones," he writes. Or, in the description of Midland-raised poet Larry D. Thomas, the wind can sand gravestones "night and day to dust." Texans have named the powerful Arctic fronts "blue northers," a term of unknown origin but of deep respect; Oklahomans call them "blue darters" or "blue blizzards."

The windiest season is spring, when winds rush through the Texas High Plains at average speeds of fourteen to seventeen miles per hour at the height of a farm windmill, though far higher sustained speeds are common. A few hundred feet above the land, the winds are even faster—and steadier, too, because no bumps or gullies or trees exist to obstruct their smooth movement. But elsewhere in Texas, the ups and downs of the land can also fuel the winds, which are often strongest on the mountain passes, where the air forces its way through, and also on the mesas, which act as a sort of tower projecting anything on them higher, into faster-moving air. "Anytime there's a change in elevation— a fairly abrupt change in elevation—you get some acceleration of the wind flow," says Nolan Clark, an Amarillo wind expert.

To harvest the wind and turn it into useful energy, inventors have tested countless designs over the centuries. There have been windmills that look like eggbeaters and windmills shaped like a double helix; windmills that look like battle-axes and those that look like busy fans. Some have had four wooden blades, like the original Halladay water-pumping design, and others, like the modern electric ones, have three slick fiberglass ones.

The towering machines of today rely on a concept known as aerodynamic lift to keep their blades turning. Planes stay in the air in much the same way as a difference in pressure between one side of the blade and the other—a difference created by the slight curvature in its long, slender shape—causes the wheel to turn as the air races past. The blade essentially swings toward the lower pressure. The amount of power in the wind is proportional to the wind speed cubed, so twice the speed means eight times the power available—which is why towers keep pushing higher, to harvest better winds. Of crucial importance, too, is capturing the wind at an optimal angle. That is why the old windmills had "tails," which would swing with the wind to allow for the maximum harvest of power.

In electric turbines, the wind turns the blades, which spin a shaft, which connects to a generator, which converts mechanical energy to electrical energy. It's not nearly that simple, of course. The modern turbines are complex machines, with at least seven thousand different components, a figure that includes tiny things like fasteners. And the technology, which has been evolving for a few hundred years, continues to change and improve, particularly in the era of electronic controls and more precise forecasting of the wind.

Each turbine in a wind farm is a lot less powerful than the giant ones used to make power from coal, which is why you need huge numbers of them to come anywhere close to the amount of electricity generated at a coal plant. "[W]hile the source of energy is certainly free," writes Richard Leslie Hills in his book on the history of wind power, "the energy itself is thinly distributed in the air because its density is low." Wind turbines face another, nearly paradoxical, challenge: the machines cannot produce power when the winds are still, and they must shut down in massive gales that could otherwise rip them apart.

The idea of harnessing the wind to create electricity goes back to 1860, when a Yankee telegraph man and chronic inventor named Moses Farmer obtained a patent after experimenting with the concept. However, aside from the electric telegraph and a few other odd inventions here and there, there was little useful purpose for electricity in that era, as wind historian Robert Righter points out in *Wind Energy in America*. So it was not until a few decades after Farmer's efforts, when Thomas Edison and Nikola Tesla were making magnificent breakthroughs in the field of electricity, that engineers began to think seriously of wind power as a source of electricity. In the late 1880s, a mustachioed Cleveland inventor named Charles Brush planted a gigantic wind wheel in the middle of his backyard, mounted atop a sixty-foot, 80,000-pound tower. He kept batteries to store the power created by the wind in his basement and fed electricity into an array of lights and electric motors. It was a marvel of its time, and it came a few years after Lord Kelvin, then merely Sir William Thomson, offered a talk in 1881 to the British Association for the Advancement of Science titled "On the Sources of Energy in Nature Available to Man for the Production of Mechanical Effect." He floated the idea that wind could be used to generate electricity, much like water at a hydroelectric plant, though he sounded a skeptical note on the cost.

It was not until after World War I, in North America, notes historian Jon Naar, "that wind-electric technology leaped forward, due mainly to the extensive wartime experience with aircraft propellers." (The relationship went the other way, too: a few decades later, when Lyndon B. Johnson was campaigning for the Senate in 1948, he had the novel idea of getting around in a helicopter, which was promptly dubbed the "Flying Windmill.") The dawn of the 1920s, Naar writes, found a number of "backyard tinkerers" pairing spare propellers with electric generators and using the resulting apparatus to turn on the lights or the radio. These "wind chargers" tended to be lighter weight than their water-pumping predecessors, by Naar's account, and often were placed atop buildings or towers.

One of these hobbyist tinkerers was a farm boy from the northern plains named Marcellus Jacobs, born in North Dakota in 1903. The family soon moved to a lonely spot in eastern Montana, and Jacobs, one of eight children, grew up relying on kerosene lamps and a gasoline generator. He was a born handyman—"when I was still in high school, I built and sold little peanut radios that operated on storage batteries," Jacobs remembered decades later—and together, Marcellus and his older brother Joseph combined the rear axle of a Model-T Ford with the fan of a water windmill and created a machine that would generate electricity. Using ideas about airplane propellers that Marcellus had learned while flying during the 1920s, they refined it to a three-bladed machine, the design that endures today. The brothers used Sitka spruce to build a few early units for their neighbors, and they started a company in Montana in 1928 to scale up their invention before moving operations to Minneapolis in 1931. There, they mass-produced it, selling about $50 million of the machines until 1956, when the company folded. A Jacobs cost roughly $1,000 in the early days, including the tower and batteries—about the same, one memoirist noted, as a new car. One of the brothers' machines, installed at the South Pole in 1933, was still whirling more than two decades later.

For farmers and ranchers, the wind machines sometimes proved a godsend, and many were glad to spend the money because being able to turn on the radio anytime made them feel less isolated. Others built their own wind chargers out of cheapness, stubbornness, or resourcefulness. Millard Holman, a blacksmith in the Panhandle town of Wellington, carved a blade out of an oak fence post, built a tail, made slip rings for the electrical connection out of copper pipe so that the wiring wouldn't get twisted when the turbine spun, and attached the contraption to a generator and voltage regulator from a wrecked automobile. He perched it atop a salvaged windmill tower, laid batteries (which he had acquired in a swap for blacksmithing work) at the tower's base, and connected it all to two lamps in the house—one in the kitchen and one in the living room—plus the family radio. Holman did "not [have] enough money to buy a 'wind charger' out of the Montgomery Ward catalog," says his grandson, Adam Holman, who has spent years as a wind tinkerer himself at West Texas A&M University. The younger Holman never got to see his grandfather's machine but heard about it from his father and other relatives.

By the 1940s tens of thousands of such wind-electric systems existed on farms across the West and Midwest, according to Robert Rodale, a journalist and organic farming specialist who wrote the foreword to Naar's *The New Wind Power*. If a farmer wanted electricity, he generally got, or built, a windmill and paired it with big, blocklike batteries encased in glass that had to be checked constantly for burnout and protected from leaves, bugs, and curious children. But just as innovators and tinkerers were changing and improving wind-electric machines, countervailing forces were gathering to prevent the technology from becoming a major supplier of electricity. In 1882, a few years before Brush used his enormous makeshift wind machine to create power, Thomas Edison wowed New York City by turning on the lights from his coal-powered Pearl Street generating station. Edison's work spread rapidly through American cities (though he, too, fantasized about using the wind to light farmsteads around the turn of the century, before Marcellus Jacobs was even born). The big cities in Texas got electric lights in the late nineteenth century—San Antonio and Dallas–Fort Worth in the early 1880s, and even Abilene, a modest settlement in the heart of West Texas, by 1891. In Austin lawmakers were so pleased with their new electric lights that they installed round chandeliers in the capitol chambers, and their bulbs spelled "Texas."

To power the lights and trolley cars of city dwellers around the country, the government turned its eye to the rushing rivers. Austin was among the first places in the nation to build a hydropower dam when it erected one on the Colorado River in the late nineteenth century. Many more dams followed around the country, as Franklin D. Roosevelt's government, anxious to beat back the Depression and spread the gift of electricity far and wide, created vast projects like the Tennessee Valley Authority or the Columbia River dams. Woody Guthrie sang about these engineering feats only a few years after leaving Pampa and joining the government payroll ("Your power is turning our darkness to dawn / Roll on, Columbia, Roll On").

Texans on isolated ranches read enviously in the papers about the lights and refrigerators and other marvels created by electricity. They were far beyond the reach of the wires: in 1935, the year of the first night baseball game in the major leagues, more than 6 million of 6.8 million farms in the United States still operated without electricity. The deficiency was even more extreme in Texas, where just 2 percent of farms got electricity. Instead, isolated Texas families supplemented their wind charger, if they had one, with the technology of another age—kerosene. Kerosene could be made from the oil that Texas abounded in, and it powered everything. Kerosene lamps made light, kerosene ran the icebox, and, mixed with a little water, it could even cover outhouse material to keep the smell down. Hal Phelps, an electrical engineer who spent his childhood around the time of World War II near Spearman, in the northernmost part of the Texas Panhandle, remembers that the lights on his family's wind charger would dim as the winds got weaker, and then the family would turn back to the old standby, kerosene. He was just learning to read. "I remember those lights going up and down as the wind would go up and down," he says.

Pres. Franklin D. Roosevelt decided he must act. Over the course of just a few decades, local utilities had established monopolies over the provision of electric power, and they had no interest in looping in America's farms. Laying lines to rural areas was too expensive, they groused, costing as much as $5,000 per mile, whereas they could make easy profits off their clusters of urban customers.

To force the utilities to get going, Roosevelt created the Rural Electrification Administration (REA) in 1935, and the next year he asked Congress to give the new agency powers to extend loans to farmers, who were busy banding together in cooperatives to ask for the lines. Fighting and dragging their feet—and prodded in the Hill Country of Central Texas by a formidable young congressman, Lyndon B. Johnson—the utilities eventually built wires to rural Texas. Farmers near Bartlett, a town fifty miles northeast of Austin, claimed to have been the first group in the nation to turn on the lights thanks to an REA loan. Three of them paid $50 each, the REA put up a $33,000 loan, the power company put up a wire, and in March 1936 one farmer was able to flick a switch and get light, after he had paid a $5 deposit for an electric meter, too. Deaf Smith County, a land of sorghum farms and cattle ranches that got wires in 1937, was the first rural part of the Panhandle to benefit. The expansion of service around Texas got delayed when World War II intervened, but the work was soon finished after the war. America would run on centralized power, striking a blow to the notion of rural self-sufficiency.

The wind chargers were suddenly obsolete, and the old water windmills, too, were often replaced by electric- or diesel-powered pumps. "After the war, windmills began disappearing from the lone prairie like the buffalo," wrote J. B. Buchanan. Even the trains switched to diesel engines, Buchanan recalled, and plenty of old Eclipse windmills by the tracks, whose name had been "as well known as the Stetson hat or the Winchester rifle" in the days when they pulled up water for steam, got dismantled for scrap during wartime. Buchanan's earliest boyhood memories involved getting spanked for climbing windmills, and when he died in 2003 at the age of ninety-six, he left behind a collection of dozens of models that still stand in a field beside the road leading into Spearman.

The utility companies would sometimes force the farmers to get rid of their wind chargers before hooking a property into the grid, presumably so they could be confident of selling the farmer the maximum amount of power. "They wouldn't tie on if your wind machine was going," recalled Wiley Stockett, who had built and sold wind chargers in the Panhandle in the 1930s, in an interview decades later with wind expert Paul Gipe. Perhaps as a sort of revenge, the Panhandle winds could blow so strongly that they'd knock out a fuse; one such instance was recorded by West Texas pioneer Nellie Witt Spikes.

Some farmers, suspicious of a modern system that took control out of their hands and entrusted it instead to a faceless utility, held out. One of these was a Panhandle man named Joseph Spinhirne. He had 400 acres west of Amarillo, and when rural electrification officials came knocking in 1940 he shooed them off. "They were very cagey, offering you the first half-mile for free, but then they wanted to charge $23 a month just to rent the line," he told Jon Naar. Another power utility proposed to charge $1,500 a mile to string a line to his farm. It was too much, Spinhirne decided. Years earlier, he had figured out how to use a wind charger to power the radio, so he turned again to his own skills. In the 1940s he bought a Jacobs wind machine for less than $100 and built the accompanying tower by himself, even as his neighbors gave up on wind electricity and upgraded to the REA. He bought their castoff appliances, like washing machines and air conditioners and power tools, and fitted them to his system, which used a different type of voltage and current from centralized power. Kenneth Starcher, who works in Canyon, not far from Spinhirne, remembers him even in the 1980s as being "modern with 1940s technology."

"Of course it helps to be a good mechanic," Spinhirne told Naar, adding, "All it takes is a lot of stubbornness and a lack of money!" It was not until the end of the 1980s, Starcher says, that Spinhirne's wife prevailed on him to hook into the grid, to indulge Nintendo-loving grandchildren.

But for every diehard like Spinhirne, thousands of others went over to the REA in the 1930s and '40s. For men like Marcellus Jacobs, the spread of electrification to farms and ranches spelled doom. Even Brush, the Cleveland inventor of the behemoth electric windmill, eventually succumbed to the luxury of electric wiring, and his machine, stored in a warehouse after his death in 1929, got destroyed in the 1950s on the orders of a manufacturing company executive who needed more space. "No wind-electric company survived" rural electrification, writes Robert Righter. Jacobs stopped making wind machines in his Minneapolis factory in 1956, and his rivals, too, vanished into history. By 1965 just 2 percent of Texas farms were without electricity—a dramatic turnaround from just a few decades before. Some kept the wind tower purely as a perch for a television antenna.

But memories of the pre-REA days, when machines fueled by the wind created power for homes, would never leave the prewar generation. Sixty miles north of Joseph Spinhirne lived another of these late adapters, a tinkerer and autodidact named Andy Marmaduke James. His son, Joe James, grew up in the Panhandle town of Dalhart, too young to remember the worst of the Dust Bowl but old enough by the time of Pearl Harbor to remember young men marching off to war. The family's modest ranch house outside town lacked electricity, but Andy Marmaduke James knew something about the wind that whistled at the eaves and sometimes forced a man to shield his face as he walked. Decades earlier, the promise of cheap land had brought him down from Indian Territory, modern-day Oklahoma. Starting at the turn of the century, chunk by chunk, Andy Marmaduke built, with his brother, one of the formidable ranches of the Panhandle. The James Ranch at one time took up most of Dallam County, in the northwest corner of Texas, and even reached into much of the Oklahoma Panhandle as well as Colorado. At the height of the ranch's operations, in 1918, it had 132 windmills, and four crews of windmillers working around the clock to keep them up. But that year, James lost nearly everything. A man had purchased the whole herd but had yet to pick them up. The brothers bought new cattle of their own to replace the herd they had just sold. Winter struck, a severe one, and almost all the cattle—about 75,000 head—froze to death.

The brothers worked the ranch back up, but they never again had the clout and riches once so firmly in hand. And so, when Joe was still a boy, his father focused on modest leaps, like bringing electricity to the small ranch house. He had only a sixth-grade education, but each night Andy Marmaduke cracked open whatever volume of *Compton's Encyclopedia* he had been reading the previous evening and proceeded to take in the next alphabetical entry. By day, he was the sort of man who took things apart and put them together again the way some people shuffle cards, out of a kind of habit. With the aid of his son, he affixed a wooden blade, probably purchased at a hardware store, to the top of a garage next to the house, about twenty feet in the air. He wired the contraption to a six-volt battery, and then to an automobile taillight that he hung in the ranch house's kitchen, or to a radio on which the family could hear the latest cattle and wheat prices via a station broadcasting out of San Antonio.

Eventually, like so many other parts of rural Texas, the property was connected to the grid. But the erection of the wind turbine, and the self-sufficiency underlying it, was a lesson Joe James would remember several decades later, when in the teeth of the national energy crisis he planted brand-new turbines, larger and more powerful than the one he and his father had put together, next to the football field of the West Texas church where he had become a priest.

CHAPTER 3

THE OIL EMBARGO

In January 1973 a wave of bitter cold and snow settled across Texas. In Lubbock, where Joe James had found a job as a priest, temperatures never rose above twenty degrees for a brutal three-day stretch. In Beeville, an oil and gas town more accustomed to muggy Gulf Coast humidity, children rushed outside to build snowmen. In Austin the streets iced over, sending cars into ditches, and fender benders were "too many to count," in the recollection of one amazed newcomer. For five days, the capital shivered in below-freezing temperatures, an anomaly for a city not used to needing thick coats.

For utilities charged with keeping people warm, these were frantic times. There was not enough natural gas flowing through the pipes to meet the spiraling demand. In Austin the University of Texas cancelled classes for a week because gas supplies were running low. To save energy, cities turned off their lights at midnight, reported *Texas Monthly* writer Paul Burka.

For Texans the sudden shortages amounted to a rude shock—a blow, even, to state pride. For more than seventy years, Texas had identified itself as the energy heartland. In 1901 a ragtag group of men— geologists, financiers, dreamers—had changed the course of history when greenish oil spurted up from a mound they had drilled into near the tiny town of Beaumont, in southeastern Texas. The mound became known as Spindletop, named for a tall tree's protrusion from a nearby hillock, and wildcatters from across the country had hopped on the first trains to Texas in the great rush to find more. Though some were less successful than others, by the 1920s rigs as far afield as the Panhandle were soon pumping fuel that would supply automobiles, sometimes faster even than pipes could be laid to carry it.

Texas oil production marched upwards as the years went by. Visions of great wealth drew dreamers and schemers like Dad Joiner, who drilled a hole in East Texas in 1930 and, at 3,536 feet, found the biggest oilfield of them all. Texas oil helped fuel ships and planes the Allies needed to defeat Nazi Germany and imperial Japan. The ready supply of crude spawned a huge network of refineries and chemical plants and other factories, which transformed the Gulf Coast into an industrial power- house. By 1945 the number of producing wells in Texas topped 100,000 for the first time, and by the mid-1960s that number had nearly doubled again. By the 1970s Texas provided a quarter of the nation's crude oil.

Natural gas, meanwhile, spent those early decades disdained as the poor cousin of oil. Some of the great fields of the Panhandle, where Michael Osborne and Joe James grew up, disappointed their discover- ers because they produced mainly gas. Natural gas could not fuel auto- mobiles, and during the first half of the twentieth century it had only limited use in home heating, because building pipelines to carry it was costly and difficult. Often, when gas came up with the oil, drillers would simply flare it off, creating a great orange glow that could be seen miles away.

After the Second World War, that began to change. During the 1950s and '60s, companies invested heavily in natural gas pipelines, using technology and labor that became available after the war. Gas was in demand, and flaring from the Texas gas fields was, in theory at least, banned during the 1950s, according to *Texas Monthly*'s Burka. Power plants began to burn gas, which was cleaner and cheaper than the oil it often replaced. By the 1970s natural gas fueled some 90 percent of Texas's power production, and nationally it supplied about half of homes with heat. Many of the water wells of the Great Plains, once powered by wind, now largely relied on natural gas engines. Even in 2012, "all the bigger pumps are natural gas, and all the smaller pumps are electric," said Nolan Clark, a Panhandle wind expert, who explained that as declining aquifer levels made water wells less productive, farmers have tended to switch to electricity.

But gas worked only when it was available. That was the lesson of that awful winter of 1972–1973, when the heat and the lights did not stay on when people huddled at home needed them most. A couple of things had gone badly wrong. The first, which didn't apply just to Texas, was that the regulatory system for gas pipelines was counterproductive. For reasons relating to individual companies' monopoly control of pipelines, more gas was being produced than ever before. In Texas, gas production peaked in 1972 and has not returned to that level since, despite even the recent shale-fracking boom—but there were still gas shortages that winter around the country. In Texas the issue was exacerbated when a major gas producer, Coastal States, could not deliver all the gas it had promised to cities like San Antonio and Austin. Later, Coastal States agreed to supply the gas, but for far higher prices. "I think the bastards ought to be put in the penitentiary," Frank Erwin, a former lawyer for Coastal States, told *Texas Monthly*'s Burka.

One man particularly anxious about the gas shortages was Max Sherman, a young state lawmaker from Amarillo. Sherman had grown up in Phillips, a tiny Panhandle town named for the Phillips Petroleum Company, an employer so dominant that if a school band member wanted a bassoon or a football player a knee brace, he got the bassoon or the brace—courtesy of Phillips. During the 1950s, Sherman recalls, gas flares were so bright that if you took a date home to a part of town called Sunset Heights, it was almost like daylight. Boys played basketball between the oil tanks. He was a bright student and earned a place at Baylor University in Waco—on a Phillips Petroleum scholarship. "I grew up in industrial socialism," Sherman likes to say.

Sherman was elected to the State Senate in 1971, and he represented a rural Panhandle district where farmers and ranchers generally liked government to stay out of their affairs. But by the early 1970s they wanted its help. Farmers began calling his office to complain that they could not get diesel for their tractors or natural gas for their irrigation pumps. Sherman could not order up a magic fix, but with the whole state suffering, he decided the way to tackle the problem was to round up the best thinkers in Texas to figure out what to do. So he sponsored legislation to create a council that would advise the governor on how to fix the energy supply problems in Texas—including, potentially, ways of getting new fuels going if oil and gas production, which began declining in Texas after 1972, could not meet the population's needs.

And so it was that in May 1973, Texas governor Dolph Briscoe, a middle-aged cattle-ranching Democrat from a South-Central Texas town called Uvalde, signed an executive order creating what he called the Governor's Energy Advisory Council. "WHEREAS, shortages of fuel and energy have already affected many thousands of Texas citizens; and WHEREAS, almost one out of every 10 Texans is directly dependent on the energy industries for his livelihood," Briscoe's memo began. He tapped his lieutenant governor, an old newspaperman named Bill Hobby Jr., as chairman of the council, thus empowering it as a political force. The task of the council, Briscoe wrote, would be to "assist the Governor in avoiding a potential energy crisis, and to make recommendations for coordinating the state's approach to energy-related problems."

Five months later, a full-fledged energy crisis arrived.

October 17, 1973, was "energy Pearl Harbor day," in the words of S. David Freeman, a future general manager of the Austin-based Lower Colorado River Authority (LCRA). On that day, Arab members of the Organization of Petroleum Exporting Countries (OPEC) declared an embargo on oil exports to countries that had backed Israel during the Arab-Israeli clash known as the Yom Kippur War. The United States used a prodigious amount of oil for its automobiles, all the more because the interstate highway system built a generation earlier made it possible for Americans to drive farther and faster than had been possible ever before. About one-third of that oil was imported, much of it from Middle Eastern countries such as Iran and Saudi Arabia. The embargo bit deep, so as 1974 dawned, the price of oil more than doubled.

For Texas oil drillers this was clearly good news, even if they knew better than to flaunt it. With world oil prices through the roof, the Texans stood to reap vast profits for producing oil that, just a few months before, had been worth far less. "The best thing that ever happened to me was the Arabs' embargo," one geologist told Roger M. Olien and Diana Davids Olien for their book *Wildcatters*. It was also good news for state universities, which controlled large swathes of land and reaped riches from leasing those lands to oil companies. The average per-acre bonus for signing a lease for University of Texas lands doubled between April and December of 1973. State government coffers, too, filled with oil money, allowing Governor Briscoe to boast of a massive surplus.

For most Texans and other ordinary Americans, however, the months and years after the embargo were nerve-wracking. Oil accounted for nearly half of the energy used by the United States. Now, not only was the amount of oil drilled in Texas, the country's biggest producer, falling, but foreign imports had become fickle, too. Gas station lines began to stretch around the block. Mack Wallace, a fiery Democratic lawyer appointed to the Texas agency that regulates oil and gas (known, curiously, as the Railroad Commission) just a month before the embargo began, recalled "being on my hands and knees begging for oil from unstable sources" at the time. "It is not an activity I wish to repeat," he told an interviewer years later.

Policy makers tried to tamp down consumption. Pres. Richard Nixon, though mired in Watergate, signed a bill slashing federal speed limits to a maximum of 55 miles per hour to save fuel. He also asked Americans to keep their wintertime thermostats at 68 degrees. "My doctor tells me that in a temperature of 66 to 68 degrees, you are really more healthy than when it is 75 to 78, if that is any comfort," Nixon told Americans in a 1973 address.

At the same time, a desperate quest for more oil ensued. Construction of the Trans-Alaska Pipeline, to tap the Prudhoe Bay oil field—North America's largest—was put on a fast track so that it could bypass environmental roadblocks. In 1975 Pres. Gerald Ford signed a law that established vast underground caverns in East Texas and Louisiana as a Strategic Petroleum Reserve, to be available in case of future crises.

Texas officials complained loudly about federal policies that kept interstate price controls on natural gas and (after 1980) taxed domestic oil, but they also echoed the call for energy conservation. In a remarkable speech a month after the embargo began, Governor Briscoe castigated the "wasteful use of energy in every segment of society." He asked that state agencies cut their use of motor fuels by a quarter, buy small cars, and set thermostats to 65 degrees in the winter, 78 degrees in the summer. Briscoe had put a brigadier general, James M. Rose, in charge of the Energy Conservation Task Force, and Rose regularly reported on how much electricity and natural gas usage at the majestic state capitol building had fallen. No detail was too small: Rose even sought to adjust the schedules of janitors at the capitol so they would not waste light at night.

But Austin and Washington had other, grander ideas about the way to reshape America's energy landscape, beyond drilling for more domestic oil and reining in Americans' insatiable appetite for fuel. Politicians took a new look at alternatives, ways to power America's lights and televisions that did not require commodities like oil and gas that were expected to run out, or the dirty coal that Texas and other states were increasingly relying on, or even the expensive and risky nuclear reactors. The term "renewable energy" was not yet part of the national lexicon, so instead policy makers began to dream, as one Houston headline writer put it, of "exotic fuel sources." Engineers called it solar energy because wind and trees and, of course, solar rays, owed their existence to the sun. Michael Osborne's first-in-Texas wind farm would even, in one rendering, be termed a "solar wind farm."

The federal government's interest in alternatives in the wake of the turbulent events of 1973 had the feel of shaking awake long-dormant dinosaurs. Electric wires had slowly but steadily marched across the nation, reaching even the most tucked-away Americans after the Second World War. Farmers in the 1950s and '60s no longer needed small-scale turbines, the type that Marcellus Jacobs had built in his Minnesota factory. Electricity had become the domain of centralized systems, feeding off of the likes of natural gas or oil or coal or, newly, nuclear power.

The turn to alternatives, particularly wind power, recalled earlier efforts to find cheaper ways of producing electricity. In the mid-1930s, as the country was pulling out of the Great Depression, a Massachusetts engineer named Palmer Putnam became irked at the high electricity bills he had to pay for his new home in Cape Cod. He joined with the S. Morgan Smith Company, an engineering firm, and together they decided to build a giant two-bladed wind turbine, far more powerful than any such machine created before. It would dwarf the huge wind wheel built in the 1880s by Charles Brush, the Cleveland inventor, and also a two-bladed machine the Soviets had erected in the Crimea in 1931. Searching for a suitable test spot, Putnam found that the general knowledge of wind conditions in the mountains was "meager and uncertain." Nonetheless, he eventually settled his 110-foot tower in the heart of Vermont's Green Mountains, on a 2,000-foot hill called Grandpa's Knob, where the local utility hoped to use the spinning turbine to throttle back its hydropower production when the wind was blowing. The first electricity flowed from the machine in 1941, but equipment failures and the war slowed things down. One night in March 1945 one of the eight-ton steel blades came loose in hurricane-force winds and flew 750 feet down a hill. Putnam's 1.25 megawatt machine—whose real failing, arguably, was that it didn't make power as cheaply as its boosters had hoped— never spun again.

Even as it disintegrated, the Smith-Putnam experiment drew careful study from a federal engineer by the name of Percy H. Thomas. Like Putnam, Thomas believed that the size of the machine was crucial, and the bigger the better. The one-to-a-farmer machines that Jacobs and his contemporaries built could never impact the nation's electricity supply in a meaningful way. But giant turbines, spread across windy areas, could make a difference. After studying the patterns of the wind, Thomas sketched designs for two enormous "aerogenerators," one capable of producing 7.5 megawatts of energy and the other 6.5 megawatts—theoretical capacities, far greater than the modern turbines in the United States today. The more modest machine would have twin three-bladed generators, each 80 feet long and both perched on top of a 475-foot tower, with the result somewhat resembling a propeller aircraft on stilts. The 7.5 megawatt one would operate with a twin set of two-bladed turbines. A budget crunch during the Korean War brought the project to a halt.

Three decades later, with oil and gas prices high and supplies threatening to run short, the federal government got serious about wind again—more serious than it had ever been before. In June 1973 the first national wind workshop took place in Washington, D.C., convened by the space agency, NASA, and the National Science Foundation. Participants watched a color movie about the Smith-Putnam turbine in Vermont and heard a detailed lecture on Percy Thomas's designs. Concerning the recent history of wind power, "it seems there was no great motivation for developing large cost-competitive wind plants," the introduction to the conference proceedings states.

The same year, 1973, the federal government dedicated money to begin funding a wind program. It began with $300,000—an amount that would rise to $67 million by 1980. Far more federal energy dollars went into jazzier priorities, like nuclear research. The federal government tapped Lockheed, better known for its work on aircraft, to look into the country's wind potential; in 1976 company researchers reported back that by 1995 fully 19 percent of U.S. electricity needs could be met with wind turbines, if $158 billion were invested.

Texas, too, began to think about the wind in a new way. Less than forty years had passed since rural electrification had transformed the lives of farmers and ranchers. State officials, many of whom hailed from the countryside, carried a certain nostalgia for that time. Bill Hobby Jr., who served as lieutenant governor during the energy crisis, recalls that one day Dolph Briscoe, the governor, "mentioned that when he was growing up on the ranch near Uvalde, the house was lit by electricity generated by windmills . . . and stored in automobile batteries in the basement." Max Sherman, the state senator from the Panhandle who had backed an advisory council to figure out how to ease the fuel shortages, remembered having grown up in a virtual "tornado alley." Though raised in the heart of oil and gas country, Sherman had come to believe that Texas would be strongest and most secure if it relied on a variety of fuel sources. "Wind was very new," he remembered nearly four decades after the energy crisis, "but at least it was something that ought to be tried."

Among the first moves of the Governor's Energy Advisory Council, dreamed up by Sherman and chaired by Hobby, was to order up a report on the state's wind potential. For expertise, the officials on the council turned to two scientists in the Panhandle—Sherman's realm—who had just realized that researching wind power would be far more interesting than their day jobs. Vaughn Nelson, a tall, understated man who had grown up in Kansas, racked up a terrible record coaching high school basketball and eventually landed a job teaching physics at West Texas State University in the small town of Canyon, yearned to be on the cutting edge: "You're teaching physics that other people have done," he said years later. "You're not doing any new stuff."

Nelson found a like-minded thinker in Earl Gilmore, a scientist at Amarillo College who had also worked at the Helium Research Center in Amarillo, a city that occasionally billed itself as the "Helium Capital of the World" and provided crucial supplies of the gas for balloons, airships, and scientific research during World War II. Gilmore, who had been born in the Panhandle town of Turkey and served in the Pacific during the war, had started daydreaming about wind power during a teaching stint at Texas Tech University. "We had some real bad sandstorms down there when we first moved down there, and I was thinking—well, maybe there would be a way to keep the velocity of the wind down by building wind turbines, by taking some of the energy out of it," Gilmore recalled years later, with a chuckle.

He approached Nelson, and they started by studying how fast the winds blew. Suddenly, they were the go-to wind experts in Texas. They went to Washington to present their ideas to that first national wind power conference in 1973. When Hobby's shop came calling the same year to ask for a report on wind potential, they were happy to help—not least because it meant a little extra cash. "We thought we were really great because we had some summer salaries," Nelson recalls. They also knew that they were in a perfect place to find wind: the wind speeds in Amarillo averaged about fourteen miles per hour at a height four times as tall as a man, and only rarely did the winds become calm.

The 1974 report Nelson and Gilmore produced for the Governor's Energy Advisory Council is called *Potential for Wind Generated Power in Texas*. It was 159 pages long and bursting with charts and maps and numbers. This was the first time Texas' government had ever funded a wind study, and the report exuded optimism: "All things considered, the region of greatest wind energy potential in the United States is located mainly in the Panhandle of Texas," the authors wrote, before adding a passing acknowledgment that Alaska might actually have the upper hand. If the state made a giant investment—something the authors acknowledged was next to impossible—Texas wind could provide "about 8 percent of the nation's anticipated demand for electrical power in 1980."

The Nelson/Gilmore report built upon the only solid data on wind speeds available at the time, which mostly came from the National Weather Service. Fifteen stations around Texas provided historical information on wind speeds, velocity, and duration. Since 1965 they had recorded weather information on magnetic tapes in three-hour intervals. But this data had problems. Most measuring stations were located at airports, and before the early 1960s they had not been placed at uniform heights, which mattered because the winds are stronger at higher altitudes. Also, some of the measuring devices, called anemometers, had sat atop buildings, but buildings block the winds and skew the measurements. (The air turbulence created by buildings is why few wind turbines appear on city roofs, even decades later. It's also why the big, modern-day turbines on the wind farms of West Texas are spaced far apart, several blade lengths from one another; otherwise, they would disrupt each other's airflow.) Getting the measurements right was crucial because even a small change in wind speed implied a very large difference in the energy a turbine could provide. For example, the report said, being off by four miles per hour on wind speed in Amarillo would imply a 69 percent difference in the potential energy. And finding the best site for the wind machine was arguably the most important decision a company would make. The World War II wind machine at Grandpa's Knob in Vermont had been a disappointment, Nelson and Gilmore noted, because wind speeds there were about 30 percent slower than its backers had anticipated.

The recommendations of the Nelson/Gilmore report were staunchly bullish. Texas, it said, should give "a significant level of support" for wind programs. The wind's fuel was free and would never run out, unlike oil and gas. Also, the machines—unlike fossil fuel or even nuclear plants—could create electricity without even needing water, a bonus in semiarid West Texas. Sure, some problems had to be solved: the wind did not blow all the time, and technology to smooth out production by storing the electricity did not exist on a large scale yet (it had not, in fact, progressed much beyond the glass batteries of Marcellus Jacobs's time). Safety and insurance would have to be considered, and some people might complain "from an aesthetic viewpoint," but the state should use the opportunity to pursue more research, the report argued, especially since the federal government was, for the first time in history, pouring money into the wind industry.

On the flatlands of North Texas, a father-son engineering team had also figured out that wind was where they should be putting money—their own, private money. By the mid-1970s the two had begun building turbines, and they dreamed of a market stretching across the world. With the government now engaged, there might be a way to grow that market.

Jay Carter Sr. and Jay Carter Jr. owned a sprawling warehouse on the edge of Burkburnett, a town that had earned the moniker "Boomtown, USA!" after oil was found nearby in 1912. Carter Sr. had been born in Burkburnett in 1923. His father, Ray Carter, worked in nearby Wichita Falls as the superintendent of Continental Oil's refinery, which was an offshoot of the Standard Oil empire. But the family moved to Wyoming and later to Tulsa, Oklahoma, as Ray Carter shifted jobs in the company. "I was an oil field brat, you might say," says Carter Sr. He tagged along as his father followed the oil fields around the country, and that got him hooked on engineering. The Carter family began moving back to Burkburnett during the mid-1960s.

The Jay Carters liked to make things—anything that had wheels and gears and seemed cutting-edge. Carter Sr., who as a child had loved building model airplanes, specialized in the engineering of lighter-weight composites, and his work on the Polaris and Minuteman rockets, both Cold War missiles, has been recognized in the Smithsonian's Air and Space Museum in Washington. When his son, Jay Jr., was five years old, his father designed a working tractor for the boy. When Jay Jr. was in high school in Ardmore, Oklahoma, his father built him a fiberglass pole for pole vaulting. When Junior won a college football scholarship to play cornerback at Texas Tech in Lubbock, he naturally became an engineer like his father, a Tech alum.

Jay Jr.'s first job was in helicopter design. In college he built two of his own gyrocopters using his father's fiberglass composites, and in 1968, when he was twenty-two, Bell Helicopter came calling. Bell made him the youngest employee in the research and development division, charged with sculpting the company's blades on its XV-15 Tilt Rotor, an aircraft that combines aspects of a helicopter and an airplane. Because he was already almost completely bald, his colleagues figured he had transferred from another job, as opposed to the kid wonder that he was, fresh out of college.

In a couple of years Jay Jr. grew restless, and he ditched the Bell Helicopter job to move to Burkburnett and rejoin his father in the town his grandfather had once virtually owned, with the mission of building a steam-powered car. Father and son stripped out and reengineered a 1968 canary yellow Volkswagen Squareback, and in 1974, after several years of soldering and pipe fitting, they showed off their invention in Ann Arbor, Michigan, impressing the assembled great and good more than the Detroit car manufacturers did. In a cover story for *Popular Science* in October of that year, a reporter wrote that Carter Jr. "feels that much of his success comes from shaking off ancient, hoary thinking about steam machines." The Carter Steamer, the reporter added, was a "tremendous achievement."

But the steam-car business was hard. Carter Jr. had paid himself no more than $250 a month, and he was looking for something that would make a buck. So he and his father settled on wind blades. The oil crisis had sent electricity prices soaring, and on the edge of Boomtown, the younger Carter had this thought about wind power: "It's an oil well that never runs dry."

They were determined to slim down the wind turbines so that they did not become the heavy giants that other companies were building. They used lessons from the helicopter industry: the lighter the turbine, the lower the capital costs to manufacture and ship. Eager to cut weight, they began with a radical premise: two blades are better than three.

3.1. *Jay Carter Jr. installing the first Carter wind turbine ever sold, in the Texas Panhandle, in the late 1970s. Courtesy Alternative Energy Institute, West Texas A&M University.*

"Some people say [the three-blader] looks better," admits Carter Jr., and, indeed, past inventors like Marcellus Jacobs had opted for three blades for extra stability. But, says Carter Jr., the three-blader is "heavy, so those who do not know any better assume it must be stronger and more robust." Carter Jr. was cocky, and he was certain the turbines would be ready in a year. But it took three to piece together the drastic redesign.

Two blades or three, the giant rotors faced problems of high winds and inefficiency if the wind direction changed. Developers also faced the more prosaic challenge of just getting the enormous things off the ground. The Carters wanted to solve these problems, and they started by orienting their turbines in exactly the opposite way recommended by handed-down wisdom. Their turbine would be lighter, smaller, and more nimble—the blades almost flexible, with a tilting hub at the top of the tower, a bit like a bobble-head doll, that would elegantly, mechanically weigh the balances of gyroscopic, centrifugal, and lifting forces to find the appropriate angle in a given gust.

To guard against overspeed, the Carters refined this method. As the wind blew faster, the blades moved to a less efficient aerodynamic angle, and therefore stalled out. Conventional turbines faced the wind the way sunflowers face the sun, with enormously large, rigid blades bearing the brunt of the wind. The Carters decided to face their turbines downwind, like a man without an umbrella, hunched over and walking backward against a hard gale. Finally and most importantly, using material from Carter Sr.'s composite work to cut weight, the rotor and the blades were light—three to five times lighter than competitors', Carter Jr. says. That cut costs, since manufacturing is partly a function of weight, and it also made the turbines easier and cheaper to put up. Anchored in place by wires, the Carter turbines could also could tip over like a kite drawn to the ground, so that men could repair the gearboxes and blades without climbing to the top.

During those three years in the late 1970s, the small prairie around the warehouse became a test pad for wind turbines, with failures akin to the repeated disasters that bedeviled the early rocket program. Turbines toppled; they exploded in flames; blades came spinning off. In one incident, a hydraulic brake wasn't powerful enough to bring a turbine to a stop in a high wind. "It kept running and burned up the brake," says Carter Jr. "Normally, that wouldn't be so bad, but hydraulic fluid was leaking, and it sprayed hot oil onto the brake disc, which, of course, was red-hot. It ignited the oil and slung it all around. The blades caught on fire. That whole machine literally burned up in the air."

But in 1978 their two-bladed turbines, rated at 25 kilowatts of capacity (about 1 or 2 percent of the potential of big modern turbines) finally performed as all the mechanical engineering equations had forecast. The father-son team was ready to put them on the market, and they figured Texas farmers and ranchers in need of electric pumping would give the new machines a try. Soon they had an eager first customer. Vaughn Nelson, the professor in the Panhandle who had co-written the state-sponsored report in 1974 on Texas wind potential, wanted one for the Alternative Energy Institute he had just started.

Nelson, of West Texas State University, had dropped most pretense of wanting to do anything other than figure out how to harness the wind. He had moved to the Panhandle to teach physics, and now the raw physicality of the place had drawn him in. There was nowhere in the world, Nelson realized, that doing research on wind power made more sense. The winds blasted through the High Plains so constantly—especially in the springtime, when they are strongest—that locals needed no fancy studies to explain that they experienced among the highest wind speeds in the nation. But lawmakers and purse-string holders wanted numbers, and in the 1974 wind study Nelson and his collaborator, Gilmore, prepared for the state, hand-drawn "energy contours"—squiggly black lines on a white map of Texas—showed that the energy potential of the Panhandle winds was about double that of areas closer to Dallas or Waco, with Austin and San Antonio even further behind.

The pair had ended their study rather self-interestedly, with a call for state funding for a wind energy institute at West Texas State in Canyon. The legislature came through, thanks largely to Max Sherman, who retired from the State Senate in 1977 to become the president of West Texas State but remained the go-to guy for the Panhandle. ("To a large extent I continued to be the senator," Sherman says.) Nelson, Gilmore, and a third man, Robert Barieau, who, like Gilmore, had also done a stint with the Panhandle's Helium Research Center, founded the Alternative Energy Institute at West Texas State in 1977. It exists to this day, funded partly by a biennial line item in place since Max Sherman's time—$90,000 from the 2011 legislature, according to the institute's Kenneth Starcher—though the university has since been renamed West Texas A&M.

3.2. *Vaughn Nelson (left), Earl Gilmore, and Robert Barieau, cofounders of the Alternative Energy Institute in Canyon. Photo by Jon Naar.*

AEI's niche, the researchers decided, would be gathering wind data, and also testing small wind turbines, which federal researchers had essentially ignored because their ambitions were focused on far larger machines. So Nelson and his crew began to turn the tiny town of Canyon into a test bed for prototype machines, like that first Carter machine, which they bought for $10,000. Sometimes they made their own wooden and fiberglass blades, and in the earliest days they commissioned three homemade turbines from a Panhandle old-timer, Wiley Stockett ("he never would build the same machine twice," according to AEI's Kenneth Starcher, who says that Stockett also built an electric motorcycle with a range of fifteen miles). As the years went on, the researchers traveled around the High Plains to climb towers and collect leftovers from other early wind projects for study. Nelson remembers dragging his long-suffering wife on side trips down bumpy country roads so he could snap photographs of turbines in all seasons and all conditions. They churned out papers and, later, textbooks, on all subjects pertaining to wind. Even years later, when he was on the verge of retirement, Nelson—a man with a dry wit and a late-in-life passion for quilting—was far more comfortable talking about wind velocities and turbine dimensions than anything to do with himself.

Twenty-five miles up the road from Canyon, and fifteen miles west of Amarillo, the hamlet of Bushland, home to a few hundred people, was also taking a close look at wind machines. For a small place, Bushland has a rich history. In the 1870s an Illinois man had dreamed up the notion of barbed wire, the invention that would transform American farming and increase the need for water windmills because it cut off cattle from creeks and springs. The new barbed wire company decided it needed a big plot of land to showcase its invention. So a sales representative, figuring the Texas Panhandle would be a promising market, bought a vast tract in 1881, added thousands of head of cattle, and ringed it with the new product. Most likely because of the shape of the livestock brand, the land became known as the Frying Pan Ranch, and in 1908 the unincorporated town of Bushland sprang up on a portion of it.

Bushland had plenty of open space, and that's what made it appealing to the federal government, which set up a research station there in 1938. The station's original purpose was to figure out how to prevent another Dust Bowl by learning more about why the soil had eroded and blown away. By the 1970s, however, this was no longer as urgent a problem, due to better land-use practices and the creation of a federal program that paid farmers to keep some of their land untilled, allowing the vegetation to hold the soil in place. New irrigation methods also helped prevent a recurrence of the Dust Bowl. Another bad drought in the 1950s—the drought that had caused the dust storms that Michael Osborne ran from in Pampa when he was a boy—had convinced farmers that dry-land farming was not reliable enough. To survive, they needed to find a way to produce enough water from the aquifers beneath the earth to allow their crops to thrive in wet years and in dry ones.

The solution of the 1950s and '60s was natural gas. At that time, natural gas was so cheap that it was free or nearly free in some parts of the Panhandle, which had huge gas supplies to begin with. Gas, coupled with better pumping technology, meant that farmers could bring up more water, from hundreds of feet down in the earth, faster than was possible with a windmill-powered well. Farming in the Texas Panhandle essentially exploded.

Then came the tremors of the early 1970s. Far from being cheap, natural gas had suddenly become expensive, and it appeared to be running out, too. The farmers that Max Sherman represented in the governmental halls of Austin could not lift water from the earth to sustain their cotton and wheat and corn because they did not have enough fuel. As desperation set in, the Texas government, which had formed the Governor's Committee on Agricultural Energy Conservation, began printing pamphlets titled "Getting the Most from Every Drop: A Checklist for Saving Farm Fuels." It had a picture of a tractor on the front and recommended simple steps like cleaning engines and inflating tires to the proper pressure.

The U.S. Department of Agriculture, which was already in charge of the Bushland station, decided to spend about $250,000 to see if its workers in this windy spot on the High Plains could solve the far-reaching problem. The man assigned to lead the new research project was a North Texas native named Nolan Clark. Clark specialized in pivot irrigation systems, machines that draw great green circles across the American heartland, easily visible to those flying over it. Pivots had been invented around 1950, but the machines had mostly not yet made it to the Texas Panhandle, even by the 1970s. Most people there still channeled the water they pumped up from the aquifers through ditches, an inefficient process known as "furrow irrigation."

When the new assignment arrived, to harness the wind to irrigate croplands, Clark was already at Bushland trying to figure out whether pivot irrigation would work in the Panhandle. The answer wasn't obvious. One problem was that while systems worked well on sandy soils in the southern part of the Panhandle, they were less effective in the clay soils to the north, where they created ruts. Moreover, the Panhandle was so windy that much of the water sprayed by the irrigators simply blew away—wasted. That meant that the natural gas used to draw it up from the aquifers was wasted, too.

When Clark took on the new project, studying whether wind could replace natural gas altogether, one of the first people he called was Vaughn Nelson, the professor in nearby Canyon. They worked out an agreement to pool resources: Nelson would get his money from the state, and Clark would get his from the federal government. The agreement drew Clark far beyond his original, irrigation-focused brief. In theory, the Canyon group would test electric wind turbines, and Clark's crew would test water windmills, but that wasn't quite how it worked out. For a while Bushland, which had more space, did most of the testing for water and electricity, while Nelson, with his physics background, delved into collection and analysis of wind data. Thus it was that Nelson's 25–kilowatt

Carter turbine, purchased in the late 1970s by the Alternative Energy Institute, ended up in Bushland.

By that time, the flatlands of Bushland, right beside the interstate that runs between Albuquerque and Oklahoma City, became a test site for all kinds of turbines. Clark's tiny team—never more than several engineers plus a couple of technicians—did research on windmill models that would eventually get distributed as far afield as an Alaskan fishing village. Sometimes two or three people a day would stop and gawk at the crazy machines spinning in the fields.

Clark's people mostly tested "vertical-axis" turbines, which spun around more or less parallel to the earth (in contrast to the big modern-day turbines, which are largely horizontal axis and spin vertically). They found out some surprising things. For example, the turbines didn't necessarily produce the highest amount of power at the highest speed. (Kenneth Starcher, of the Alternative Energy Institute, likens this to the way automobiles are most efficient not at 145 miles per hour, but at 45 or 55.) They also found a couple of problems. The biggest one, as Clark saw it, was that the turbines tended to start falling apart after a year. "We found that a lot of wind turbines operated pretty good for the first 3,000 to 4,000 hours," he says. But once they hit 7,000 to 8,000 hours, "things started happening to them." Blades might fall off, generators could fail, gearboxes could fail. "If you buy a car, the bearings and wheels are designed to last about 100,000 miles," Clark says. But turbines do not carry humans and so were not designed to be nearly so hardy. Another problem was that even if the wind machines could operate smoothly, they often had trouble shutting down. That's a necessary safety precaution; even today, modern wind machines shut down when wind speeds rise above about fifty-five miles per hour so they don't get damaged—or injure anyone nearby. Turbines had to be able to shut down at other times, too, in case they needed maintenance or a tornado was coming.

The research into mechanical water-pumping machines—Clark's original assignment—was in some ways a bust. Electric windmills would not work for large-scale crops, he found, and the most fundamental problem was not power but time. Crops need water only four to five months of the year, including part of the summer when the winds are weakest. The rest of the time, the big, expensive machines sit essentially useless. "Yes, I'm probably disappointed," said Clark, reflecting on the failure of his research dream thirty years later. The technology is ready, he said, but the economics just never quite worked out.

3.3. *A vertical-axis turbine spinning at the USDA site in Bushland, a site in the Texas Panhandle that tested early-stage designs. Courtesy Alternative Energy Institute, West Texas A&M University.*

Like Vaughn Nelson and Nolan Clark, Joe James, the Dalhart boy whose father had built a wind-electric generator to create light for the family in the days before rural electrification, was also trying to make turbines spin in the powerful winds of West Texas as the nation's energy crisis intensified. He wasn't in the research business, though; he wanted to put the turbines to use for God.

James had first heard the call of the Lord as a fifth grader. During most summer months he helped his father at their ranch. But that particular summer in the early 1940s he had broken his ankle—he can't remember how—and his parents sent him to Bible camp. At recess, during a game of baseball, he was out in right field, on crutches, when a pastor came to bat. James heard a voice: "'The Lord tells me He loves me and He loves you,'" James remembers. "That call has never left me."

By the 1970s James had ascended the ranks of the Catholic Church, and in 1977 he was named to lead a congregation in Lubbock that needed a new church. He was forty-five years old, but he still remembered the lessons of his boyhood. "We're supposed to be custodians of

3.4. *Father Joe James, seen here in 2011, erected turbines in the early 1980s near his Lubbock church. Photo by David Bowman.*

God's creation," he said, years later, "but churches are among the greatest energy sieves of any buildings."

Working with a church architect, Deacon Leroy Behnke, James settled upon something near the opposite of typical soaring, high-windowed churches. His building, Saint John Neumann, would be dug out of the earth, like the homes of the first white settlers. "It would be cool in the summertime and warm in the winter," he thought. "Christianity has been here for nineteen centuries without air conditioning and heating." And remembering how his father had brought light to the old ranch house, he decided that much of the power would come from the winds whipping through Lubbock.

Putting a wind charger atop a garage was a lot easier than finding five turbines to power a church and school. James met Coy Harris, who ran an engineering company in town that had been experimenting with wind turbines. Harris said he could build the turbines for $84,000, and James set about collecting the money: the federal government, as James remembered several decades later, agreed to pay half via grants channeled through the state, and the parish would pay the other half. But Harris, who now runs a wind-power museum in Lubbock, decided he could not produce reliable turbines: "You don't sell things you're developing, especially wind turbines, because they tend to fall apart," he says. Suddenly James looked like he would have to return the government's investment. James himself, relying on money he got from renting the family ranch, says he lost a $20,000 investment he made in Harris's company. "It got pretty rough when he couldn't produce the generators," James says.

But he learned of the father-son team in Burkburnett that was already building turbines. He gave them a call, and one day in early 1982, the delivery arrived. It was just a few months after Michael Osborne, working 200 miles to the north in Pampa, had turned on the first wind farm in Texas. The wind was so sharp around Lubbock—sharp enough that kids sometimes tied sheets to their wrists, as sails, when they rode bicycles—that the Carter workers had to sand the two-blade machine to reduce efficiency, for fear the turbines would spin too fast and the whole thing would wobble like an out-of-kilter ceiling fan.

James had spent weeks checking the wind currents. He made a kite by nailing together a small wooden cross, gluing paper across it, and on the long tail of twine tying streamers every five feet. Then, like a priestly Ben Franklin, he flew the kite over the property, figuring out the best site for the turbines. They needed to have plenty of space around them so the top of the blades wouldn't hit anything if they fell, and they had to be far enough apart so they wouldn't interfere with each other's wind. "If you've got a ten-foot object, it will cause turbulence forty feet downwind," says James, in 2011 a still-vigorous seventy-nine-year-old with bushy eyebrows and a penchant for navy blue denim coveralls that make him look like an auto mechanic. The only tip-off that he was a monsignor was the heavy talisman of Saint Benedict on a gold chain around his neck.

Finally, he settled on putting the turbines beside the church school's football field.

The turbines were taller than anything else around, and James planted three of them, each sixty feet tall, along a sideline and a fourth

3.5. *Turbines spin beside the Saint John Neumann church in Lubbock. Photo by Michael Osborne.*

behind an end zone. They were wired directly into the church and the school, and in the evenings, when the lights were out and the air-conditioning was turned off, excess power could be fed to the city's electric grid. The last turbine went in next to the church; it was called Big Bird, because it stood eighty feet tall. The turbines generated enough electricity to cover a quarter of the needs of the school and the 850-strong congregation. "Everybody going down the loop could see our wind generators for a mile away," James remembers.

James's tenure was marked by one major success—or controversy, depending how you look at it. In the summer of 1988 several of his parishioners, including a retired air force man and a housewife, said they were getting messages from the Virgin Mary. Word spread, as it is wont to do, and roughly 22,000 people by James's estimate (13,000 according to local officials) showed up on an August day for an outdoor mass to celebrate the Feast of the Assumption. Some churchgoers claimed to see the sun "dancing." Several claimed to be healed. One boy was suffering from muscular dystrophy, which "'caused him to have his hand against his chest,'" James told the *Lubbock Avalanche-Journal*. "'When the Blessed Mother came to the fountain at St. John Neumann, he extended out his hand.'" The Lubbock bishop shunned the event and later appointed a "team of experts to investigate whether any miracles had occurred," according to the paper. The panel found no evidence of miracles, and in 1990 James was quietly relieved of his post.

The wind turbines stayed up, for at least a short while, but were eventually dismantled; two decades later, all that remains of the turbines are five eight-foot-deep cement anchors, sunk like gravestones around the church grounds. "The priest who followed me couldn't give a flip about them," James says. "He said he couldn't get anyone to repair them or cobble them together. If you want to do something badly enough, you will. If you don't want to, the way is filled with excuses."

CHAPTER 4

THE 1980S

BOOM—THEN BUST

When Michael Osborne drove back to his hometown of Pampa in 1981 to plant five wind turbines made by Jay Carter Sr. and Jay Carter Jr. atop waving grasses on his cousin's land and declare it the "second-largest wind farm in the known universe," he had every reason to be bullish about the industry he was helping to create. No large-scale wind farms—or wind ranches, as they were sometimes called in those early days—yet existed anywhere in the country. Osborne's project could power only several dozen homes at best, so it wasn't going to change the world either.

But it was a start, and the blustery winds of the Panhandle had a way of making believers out of people. In the fifteen-minute video Osborne made to document his creation, this optimism, along with a dose of just-do-it Texan bravado, was on display. "You're in the pay almost every day, out of the 365 days a year," declares Carl Kennedy, the Gray County judge and Osborne's cousin, who had lent his land for the project. In the film he is wearing a trucker cap, and tree branches bob up and down in the background as the two-bladed Carter turbines buzz like a helicopter preparing for lift-off. "If there's any truth to the energy shortage—and the federal government is encouraging people to develop secondary energy sources—here's a first step to getting that done," Kennedy adds.

Osborne, squinting through slightly oversized glasses while the wind riffles through his hair, is businesslike but clearly excited. "We're taking a production approach to this renewable energy," he says, noting that he is paying his cousin a royalty. "It's not just some sort of exotic energy."

Father Joe James uses his airtime to argue earnestly for the soul of humanity. It's "kind of an American idea—I can waste all I want to waste. All I have to do is be able to pay for it," James tells the camera as his turbines twirl like batons above the buildings of Saint John Neumann. "However, I don't feel as though we are free to waste. We are responsible, whether it be inexpensive or expensive. And our wind machines, for example, or the way we built the church, underground and at additional expense—they say it's not cost-effective. Well, the cost of utilities is going to continue to go up. Right now, it may not be cost-effective. Later on it may be. But is cost-effectiveness the only reason why we exist? The only thing that dictates what we use and what we don't use? I say no!"

James and Osborne felt they could see the future more clearly than the stodgy utility companies or the big corporations that liked large, expensive, polluting power projects. They would lead the state and the country onto a better path, a path that would help free Americans of dependence on dirty and costly substances buried in the ground. James remembered words his father had told him during his childhood in Dalhart: "Don't you ever touch a thing unless it's better for you touching it." That was the principle he wanted to live by.

The federal government was also trying to make things better by showering this emerging, barely tested source of energy with money. The timid $300,000 investment in 1973 had turned into a bouquet of incentives. By the time Osborne opened his wind farm in the fall of 1981, federal dollars were cascading into the wind industry. That was the year that funding for wind research and development peaked at some $186 million (in 2011 dollars), as Janet L. Sawin, later a senior fellow at the Worldwatch Institute in Washington, documented in 2001 in her outstanding 662-page PhD dissertation on wind energy for the Fletcher School at Tufts University. Big engineering companies like Boeing and General Electric soaked up the money to fund their designs for prototype turbines that grew bigger and bigger. "Make no mistake about it, wind farming is a big, big business," trumpeted a 1982 *Wind Power Digest* article, which cited estimates that the market for wind farms and big wind systems would rise as high as $150 billion.

More relevant to Osborne and Joe James and the Carters were tax incentives that encouraged everyday Americans to put up the newfangled whirligigs. In 1978 President Carter had signed the National Energy Act, a major piece of legislation that included tax credits for homeowners and businesses adding renewable energy installations. In 1980, the year he was voted out of office in a landslide, Carter signed another law called the Crude Oil Windfall Profits Tax Act, which bumped the credits up further so that homeowners installing renewable-energy systems could get a 40 percent credit and businesses could get a 15 percent credit that would run through 1985. (The law's signature provision, the tax on domestic oil production, caused a loud outcry from oil-rich Texas.) Business projects could thus reap a total credit of 25 percent after factoring in another existing incentive, so Osborne, who paid $80,000 for the five Carter machines, could cut his tax bill by a cool $20,000 because of the feds. Saint John Neumann, the church where Father Joe James put up his turbines in 1982, also got a tax credit, and, James says, was able to pay off in three years the $35,000 it had borrowed to help pay for the turbines.

Quite apart from the mesh of tax incentives, Michael Osborne and Joe James also owed their projects to additional, highly technical federal legislation that profoundly shook up the humdrum world of electric utilities. Before 1978 the utilities that created power from coal or gas or oil felt no need to buy power from small, enterprising producers who wanted to sell it to them. Why should they buy what they could make more cheaply, and with a lot less hassle, on their own? The 1978 legislation, called the Public Utility Regulatory Policies Act (part of the National Energy Act), would become arguably the most vital piece of legislation in the history of wind power. The law aimed to encourage more development of renewable energy. PURPA took a shot at the monopoly utilities, which were accustomed to producing electric power from huge, centralized plants that burned fossil fuels. Under the new law, utilities were required to buy electricity from smaller, independent sources, insofar as they existed, for a fair price.

Osborne, with his handful of two-bladed turbines planted in the grasslands of Pampa, certainly qualified as a smaller source. He was eager to test out this law because he wanted to prove he could make money from the winds that had once carried his high school glider out of sight. "It's kind of fun to get a check from the utility company," Osborne told a Dallas newspaper reporter in 1982. The check he received, for 2.69 cents per kilowatt-hour, wasn't for quite as much as he would have liked, but it was something. Before the law, utilities would have shunned projects like the Pampa wind farm and paid nothing at all. With the law in place, the experiment was a very modest success.

After he grew weary of sending young musicians-turned-mechanics out to change the oil and fix the equipment each time it was struck by lightning, he shut the turbines off in 1985 and sold them for a nominal sum to Vaughn Nelson's Alternative Energy Institute in Canyon, which was glad to have them.

As wind turbines inched their way across the land in the late 1970s and early 1980s, one ranch or fledgling wind farm at a time, the problems with the technology became increasingly obvious to would-be buyers or investors. At best the machines needed plenty of maintenance; Joe James, the Lubbock priest, said he had to winch down his tallest turbine, Big Bird, every few weeks in the summer because insects flying toward the lights of nearby buildings would end up splattered on the blades, which reduced the machine's efficiency.

At worst, the blades would fly off in a storm or break, as they had at the first wind farm in the United States, Crotched Mountain in New Hampshire. Joe James had some of these problems, too: once, he says, some of his Carter blades broke off after they got blown into the turbine pole by a dust devil, a kind of mini-tornado common to Lubbock. The blades were insured, but James says the insurance company refused to keep covering them after the incident.

That the turbines of this era had problems is hardly surprising; throughout history, young technologies have shown themselves to be clunky at best, and hazardous at worst. Texans knew this well, from the oil fields. In the frenzied decades that followed Spindletop, drill bits were notorious for jamming, and the wooden derricks, operated by men who sometimes put money above safety, could easily catch fire and entomb workers.

As part of a manufacturing outfit, Jay Carter Jr. took the early turbine failures in stride. "Anybody that thinks that you're not gonna have problems when you get into a new technology is really being kind of naïve," he tells the camera during Osborne's 1980s movie, outfitted in aviator sunglasses and a yellow windbreaker. "So we expect, and we expected, to have some problems." The Carters weren't the only ones dealing with glitches. Plenty of other companies sensed the promise of a new market and were flooding in, too, so that even a few oil rigs out in the Gulf of Mexico installed tiny wind turbines for power.

Some of the arrivals on the wind power–manufacturing scene had experience. These included Marcellus Jacobs, the inventor who had sold millions of dollars worth of electric turbines before rural electrification wiped him out. After the oil embargo he jumped back into the business, this time with his son Paul. They opened a Minnesota factory in 1980 and made machines three times as powerful as the ones Jacobs had invented nearly sixty years earlier, when he had wanted juice for his radio. Some farmers and ranchers also got busy fixing old, pre–rural electrification machines. "During the rebirth of wind energy systems in the '70s and '80s, many Jacobs units were literally dug out of the ground, cleaned, remounted on towers, and immediately began producing power," reported *Wind Power Digest* upon Jacobs's death in 1985.

But plenty of the new entrants were young men with considerably less experience than either the Carters or the Jacobses, and the results were, many would say, predictable. One of these was Carlos Gottfried, who had emerged from engineering school at Southern Methodist University (SMU) in Dallas convinced that he had figured out a clever way to build wind turbines. Born in Mexico City, Gottfried got sent to a military school in Missouri after being asked to leave another school back home ("I was out of control," he says now). He loved the new school, which shaped him up, and classmates from Texas taught him to say "y'all" and to pronounce "oil" as "ahl."

Gottfried remembered once driving with his father past a wind turbine in the mountains of Mexico, and they dreamed about how the machines could bring power to people in their own country who lacked it. But Gottfried also knew how often the turbines broke down—and at SMU, he resolved to build a "direct-drive" generator, without many of the pesky parts that gave wind farm operators so much trouble. "The idea was to build a generator that wouldn't need maintenance for twenty years," he said years later. "No gearbox, no belt, no chains."

With backing from his father and uncle, both retired U.S. Air Force pilots who had founded an electrical conglomerate in Mexico City, Gottfried set up Hummingbird Windpower in the late 1970s. Its headquarters were in Mexico City, an import-export arm was in Houston, and its manufacturing facilities occupied an old airplane hangar in Sweetwater, where female pilots who flew military transport planes had been based during World War II. Ranchers, eager to generate their own power as electricity prices headed skywards, got in line for Hummingbird machines. Gottfried says his first sale came shortly after a renewable-energy fair in Denver, when he was invited on a syndicated radio show to talk about the Hummingbird. "There was a guy who called us the next day, who got in his pickup truck and drove 300 miles to call us from the nearest phone," Gottfried says, his memory perhaps a bit exaggerated. "He told us he wanted one right away. That was our first sale."

But for all the headiness and energy in the business, the challenges began soon enough. "We had problems with just about everything," he says. "We had problems with the blades, we had problems with the controls. But the biggest problem was the inverters." Hummingbird had several hundred machines around the United States, and each time an inverter—made by suppliers in other states, according to Gottfried—failed, a time-consuming chain of events ensued. Someone would have to go out to the (undoubtedly frustrated) customer, unhook the inverter from the wall, send it back to the supplier to get it fixed, and reinstall it. That process could take several months, Gottfried says—and then "it would fail again afterwards." The situation was, he says, "a real disaster. You can imagine trying to service these customers." Competitor companies had problems as well; it was a time when "every imaginable type of problem was happening throughout the industry," Gottfried remembers.

One Hummingbird turbine, installed near the Sweetwater factory, malfunctioned during a bad storm, according to the recollection of one former employee, Caroline Crimm, who says, "The mechanical arm that turned the blades out of the wind failed, and the runaway blades came apart with a shrieking that was heard miles away." Gottfried does not remember this incident, but says, "I'm sure it did [happen]." The company did experience "runaways," he said, and "it's quite a horrific event."

After only a year or two of making and selling wind turbines, the Hummingbird manufacturing plant in Sweetwater gave up and closed its doors. In the final analysis, "We were just too far ahead of the curve" on design, Gottfried argues, noting that some large modern turbines have direct-drive technology, as his did, which eliminates the gearbox and allows for a lower-speed generator. Closing up shop in Sweetwater "was actually a relief," he says, and it forced him to go back to the drawing board to fix the problems. Now the head of a Mexican company called Potencia Industrial, he is still making Hummingbirds—thirty to fifty units a year, he estimates—and selling them around the world, in addition to a variety of other electrical equipment. In 2011 a new Hummingbird arrived at the Alternative Energy Institute in Canyon for testing—just as one of its predecessors had thirty years earlier.

The mechanical faults plaguing the industry were compounded by another challenge: the turbines had to withstand the constant weather extremes. On the High Plains, they must endure 100-degree temperatures in the summer and bitter freezes in the winter, storms year-round that bring lightning and snow or hail and, of course, furious winds.

"Tapping the wind is no easy task. It's hard as hell on a machine," Russel Smith, the executive director of the Texas Renewable Energy Industries Association (TREIA), told the *Texas Observer* in 1984. Texas would learn this lesson again and again over the decades that followed, to the industry's detriment. Wind turbines are tall and extremely visible for miles around, and while that is good publicity for a swiftly spinning turbine, it also means passersby notice and tut-tut when it has stopped.

The Alternative Energy Institute, because it was working with prototype turbines, almost inevitably had more problems than just about anyone else. Researchers initially settled in to test their turbines at a campus-owned pig farm. But it was a seven-mile drive away, and eventually university officials let them use an open field on the far east side of Canyon, adjacent to the university.

One day in the 1980s researchers began testing an Ohio-made turbine in winds of up to thirty miles per hour. These were not unusual winds for the Panhandle, and the manufacturer's representative, confident in his machine, had returned home a day earlier, after tests at twelve to fifteen miles per hour had shown no problems. But the turbine's blades began oscillating in a fishtail-type motion, meaning that each blade took a different angle to the wind, according to Kenneth Starcher, a chatty man raised east of Lubbock who began as a student of Vaughn Nelson's and rose to become the assistant director of the institute. Operators turned off the power and tried to apply the brake, but the turbine "did not slow down and was then in runaway mode," says Starcher. Police arrived to shut down a nearby highway in case the turbine slung its blades, and they kept it closed until the machine collapsed. The brake, as it turned out, had melted because the turbine had been going so fast. The test site was soon moved to another part of town, on the north side of campus. To the researchers' chagrin, it was slightly downhill—and downwind—from campus buildings, which complicated the wind patterns and forced them to put up higher towers to get good winds. "It was the crappiest wind site we could get to but still be on university property," says Starcher.

Technology was not the only source of problems. Even experienced wind hands could make errors working with the machines, especially in an age when safety rules were far less stringent than they are today. In the 1980s Starcher and another AEI researcher were out in a field doing tests on Carter machines for a project for the National Renewable Energy Laboratory. After failing to communicate properly with one another, the two detached the wires anchoring one machine to a truck and the ground at exactly the same time, leaving the turbine unanchored and teetering. Starcher almost got one of the wires back into place, but a gust of wind—he estimates about forty miles per hour—ripped the wire out of his hand. "I've got nine scars 'til this day," he says.

As the turbine toppled, Starcher's colleague, Forrest "Woody" Stoddard, leapt out of the way, dropping a brand-new Nikon camera. "My God, the thing just missed crushing him," says Starcher. "It killed the camera. He was pissed about the camera." Rattled, the researchers soon made another avoidable mistake: a burst of wind made a wire tethering a second Carter turbine to their truck go slack and jump off its pulley, as Starcher drove the truck forward. And so another Carter turbine toppled. "The nastiest sound you ever heard is a guy wire separating under tension," says Starcher, referring to the wire holding the machine in place. The second accident would not have happened, Starcher said, if the researchers had had another person present to stabilize the turbine or if they had waited until the winds calmed and it was safe to take it down, but they were too anxious to retrieve their data.

Ultimately, no one was hurt, but some equipment and egos were crushed. It was an expensive mistake, says Nelson, and he was relieved that AEI could buy Michael Osborne's Pampa turbines and salvage them for parts to rebuild the Carter turbines.

As for Carter Jr., he claimed that with proper care his turbines were ready for anything the weather could hurl at them. "We've had some units that have been through winds over 120 miles per hour and continue to generate electricity at winds over 100 miles per hour without any failures," he said in Osborne's early-1980s video. "We've had lightning strikes, we've been through marble-size hail. We've been through a number of thunderstorms and overspeed conditions—and without any related failures."

But the Carters' machines had troubles in a more mundane way—with the gearboxes. These boxes speed up the rotations created by the wind in preparation for the electrical generator. They are among the most critical pieces of a turbine, yet even in 2011, they "are really the Achilles' heel of the system," according to José Zayas, a technology expert with Sandia National Laboratories in New Mexico. Carter Jr., in Osborne's film, admits to a quality control issue: "We do have some problems with the gearbox. Gearboxes have been around for a long time. We shouldn't have had any problems with the gearbox," he said. Basically, it was tough to ensure that every part of a machine was working, Carter Jr. said, and the learning curve was steep. "We do a lot more testing of our electronics now before we ever send a unit out than we ever did before," he says in the 1982 video. "And we just didn't realize that we had to do that kind of quality control testing in order to ensure you have a real reliable product. All you've got to do is just have one little component go down, and it's only one little component in 1,000 components, and yet it shuts the whole machine down."

But machinery, weather, and human error were not the only troubles confronting Texas windmen in the early 1980s. They also faced a far more powerful foe: economics.

In the 1970s, in the wake of first the Arab oil embargo and then the Iranian revolution, law-abiding people sometimes had to wait for gasoline in lines that stretched around the block; others grew so desperate they stole it. Even as supply stabilized in the early 1980s, crude oil prices still topped $100 a barrel (in 2009 dollars)— essentially double their level in 1979, the year of the Iranian revolution.

Slowly but steadily, however, energy prices began to fall. Demand for gasoline had faltered; Americans, the most energy-guzzling people on earth, had finally figured out how to cut back. They bought more-fuel-efficient cars, under the government's exhortations, and drove more slowly. They learned to turn off unnecessary lights. Some began buying more-energy-saving refrigerators, thanks to national appliance-efficiency requirements that came into effect in the 1980s. The amount of available energy also increased. Oil supplies got a boost from the 1977 opening of the 800-mile, $8 billion Trans-Alaska Pipeline, built in just two years, which ferried crude from Alaska's North Slope to a southerly terminal in Valdez. American and European companies, desperate for ways to undercut the new power of OPEC, began searching for oil more

aggressively in non-OPEC parts of the world, like Mexico and the North
Sea. The government began an effort to promote "gasohol," a mix of
petroleum and ethanol.

The shortages of natural gas eased, too. For this Texans give partial
credit to the deregulation of the natural gas industry, which began in
1978 with the passage of the Natural Gas Policy Act, which increased
national price ceilings on natural gas. Ronald Reagan's administration in the 1980s continued with deregulation, and, ultimately, higher
prices for gas encouraged companies to search for more of the stuff.
They found it, and that in turn helped lower prices. In Texas natural gas
production, which had fallen sharply during the 1970s, essentially flat-
tened out after 1983.

Just as important, other types of energy were on the rise. The natural gas shortages of the 1970s had forced utilities to turn to alternatives
so that they could keep the electric grid working and avoid the wrath of
customers and regulators. In 1975 oil and gas regulators at the Railroad
Commission of Texas even ordered utilities to stop building new natural
gas power plants, and Congress issued the same orders for the nation in
1978 (the Texas order was repealed in 1979, and the federal one in 1987).
Alternatives, for the moment, did not mean wind or solar, because these
technologies were not nearly far enough along to run a power plant. Instead, they largely meant coal. Several huge coal plants, including the
ones that environmentalists today condemn as excessively "dirty," got
built in Texas during the late 1970s and early 1980s. (For comparison,
Osborne's wind farm, a mere 125 kilowatts, had about 1/5000th of the
capacity of the relatively modest coal plant that began operating near
Victoria, Texas, in 1980.) This was also the era when work began on
Texas's two nuclear plants.

For ordinary Americans, this proliferation of energy sources, and
especially the plunge in oil prices, was good news. They could fill up at
the pump without worrying about lines or spending their entire pay-
check on gasoline. One Exxon station north of Austin, operated by a
man with a flair for gimmickry named Billy Jack Mason, offered free
gas—zero cents a gallon—on one memorable April day in 1986. "Some
people damn near try to kill you to get in line," a taxi driver named
Michael Johnson told the *Washington Post*. "They do whatever it takes,
then they flip you the finger." Johnson had driven thirty miles to line
up at 6:00 a.m., and three hours later the access road to Interstate
35 was clogged for six miles with waiting cars. Some drivers saw it as

recompense for the frighteningly high oil prices of a few years earlier. "How many times in your life do you get free gas?" another customer, a painter, explained to the *Post*. "It's the opportunity of a lifetime."

But for most of the Texas windmen, the opportunities had ended. When Osborne dismantled his wind farm in 1985, nothing bigger, or better, arose to take its place. The government, and the taxpayers who funded it, had well-nigh forgotten the years of crisis by 1986, the year of Billy Jack Mason's free gasoline, when U.S. oil prices plunged to one-third of their levels of five or six years earlier. With the energy shortages seemingly resolved, enthusiasm for renewable energy collapsed—as, more importantly, did the federal tax incentives that had almost single-handedly propped up wind power.

A change of administration in Washington contributed to the doom enveloping alternatives. Carter was committed to renewables right up to the end. In 1979, the year he put thirty-two solar panels atop the White House roof, he declared that 20 percent of U.S. electricity should come from renewable sources by the year 2000, and in September 1980 he signed the Wind Energy Systems Act, intended to help wind projects find research and development funds. But Carter lost to Ronald Reagan two months later, and the former California governor took charge with a broad mandate and a small-government-is-better mantra.

Reagan had signaled his deep skepticism of government spending on energy even before the election. "The Department of Energy has a multibillion-dollar budget, in excess of $10 billion," Reagan said in a debate with Carter in late October 1980. "It hasn't produced a quart of oil or a lump of coal or anything else in the line of energy."

It certainly hadn't produced much by way of wind energy, either. One of the oddities of the wind business is that the modern turbines of today are not descendants of the enormous experimental turbines that heavyweights like General Electric and Boeing and Alcoa and Westinghouse produced in the late 1970s, using millions of federal dollars. Those had experienced major technical problems and flopped. Boeing struggled with dirt getting into hydraulic fluid. Alcoa, the aluminum giant, pulled out of the wind business soon after its solitary 500-kilowatt test turbine, shaped like a kitchen beater and erected in California's San Gorgonio Pass, slung a blade at one of the wires holding it in place just a few hours after being turned on. Making matters worse, this occurred just before a high-profile wind conference featuring California governor Jerry Brown was due to convene. "I have some good news and some bad news," Paul Vogsburgh of Alcoa announced to those assembled. "The bad news is that our wind turbine destroyed itself. The good news is that we did not have to evacuate Los Angeles." As the chronicler Paul Gipe notes, this was a reference to Three Mile Island, the partial meltdown that had occurred in Pennsylvania just two years earlier—and a not-so-subtle reminder that wind, for all its faults, would not obliterate entire cities.

"It's kind of strange," says Vaughn Nelson, the retired director of the Alternative Energy Institute in Canyon. "The tract of development that led to the large megawatt machines today came from what we'd call the ground up of the small machines getting bigger [with] economies of scale, rather than starting with great big machines funded by government." In 1982, the year that Nelson hosted the unexpectedly well-attended annual conference of the American Wind Energy Association, which was held in Amarillo, and the year that Houston hosted a major conference of solar-energy advocates, federal spending on wind research and development projects got slashed by more than half and kept falling, bottoming out at $8 million in 1988. But Texas companies mostly weren't reaping that R&D bounty anyway, so what really stung was the expiration of the 25 percent federal tax credits for alternative-energy investment at the end of 1985. Without the tax credit, wind power became simply a nonstarter for people with big dreams, like Osborne. "Reagan killed the industry," says Michael Osborne, his voice still tense and angry some thirty years later. Renewable energy, he said, would have been able to endure through one Reagan term, but it couldn't outlast two, because Reagan had no intention of reinstating the tax credit. Symbolically, in 1986 Reagan took down the thirty-two solar panels that Jimmy Carter had placed atop the White House roof.

There was only one place in the country that stood tall in the face of increasing federal indifference and insisted that developers were welcome to generate power from the winds that flowed across its hills. Ironically enough, it was Ronald Reagan's home state of California.

The state that ended the 1980s with about 90 percent of the wind-power capacity in the entire world does not actually have great winds. Far from it: when all fifty states' "wind resources"

are considered, meaning how strongly the breeze blows at a height of eighty meters, California places only nineteenth. The Golden State has few vast, treeless expanses like the Panhandle of top-ranked Texas. Instead, it has mountains and valleys and rolling hills, which slow or block the winds that roll in from the coast (and are especially strong during spring and summer).

But California has other advantages, and it was these that attracted swarms of wind developers, some scrupulous and some not so scrupulous, during the 1980s. Then and now, it is the most populous state in the union, meaning that there is no shortage of people who need electricity, in contrast to North Dakota or even West Texas, where wind farms can be hundreds of miles from large cities. California's electricity prices have long been considerably higher than the national average, which amounts to a built-in incentive for alternative energy. (California today remains the leader in solar-power development nationwide.) The state also has a famous, or infamous, environmentalist bent that made it dig in against nuclear plants, even in the 1960s and '70s. It was this environmentalism, coupled with the sharp jolt of the 1970s' oil crises, that swung California into action on wind.

California's wind-farm spree of the 1980s had its roots in Reagan's final years as governor. In 1974, spooked by OPEC, California lawmakers passed (and Reagan signed) the Warren-Alquist State Energy Resources Conservation and Development Act, which encouraged energy conservation. It also, crucially, created the California Energy Commission (CEC), a regulatory body given wide authority over energy planning in the state, including the power to oversee power plant siting and to approve utilities' energy planning processes.

The CEC began operating the same year, 1975, as a young Democrat named Edmund G. ("Jerry") Brown Jr. succeeded Reagan as governor. Brown, soon nicknamed "Woodchips and Windmills," proved eager to promote renewable energy and move away from oil and gas, which had been supplying 70 percent of the state's electric production. Energy at the time was a huge political issue for California. The son of a former governor, Brown arrived in office confronted with a state whose population and appetite for electricity were exploding, even as the federal government forbade the building of new gas-fired power generators for fear of exacerbating the shortages.

The obvious solution was nuclear plants, which states from Texas to Maine were building. California's utilities dreamed of putting a string of reactors along the coastline. But Brown distrusted nuclear power, even before the 1979 accident at Three Mile Island. So via the energy commission, whose members he appointed, he pursued two key policies. One was energy efficiency, meaning ways to make the state use energy more wisely. After listening to a Lawrence Berkeley National Laboratory academic named Art Rosenfeld, the governor was persuaded that just by doing a simple thing like requiring refrigerators to operate more efficiently, the state could save as much energy as would be produced by one proposed, highly controversial, nuclear plant.

The other solution Brown determinedly pursued was renewables. Brown's energy commission ordered up studies of the state's wind—where it blew and where it blew the strongest, much as Nelson and Gilmore had done in Texas several years earlier. Around the early 1980s the CEC also sought a demonstration project, and thus it was that a solitary wind turbine, made by Jay Carter Sr. and Jay Carter Jr. of Burkburnett, Texas, ended up standing atop a hill along Interstate 80 between San Francisco and Sacramento, the capital—a "well-traveled political highway," Jay Jr. says—where state officials could gaze out from their automobiles and ponder their state's energy future.

The California wind rush was soon on, full tilt. Many of the state's best winds could be found in mountain passes—not the tall, snow-capped Sierras, to be sure, but lower and more accessible rises. It was in these rises that the state's trio of major wind farm sites—San Gorgonio east of Los Angeles, Tehachapi north of Los Angeles, and Altamont Pass, an hour's drive east of San Francisco—got built. By the end of 1981 California had 10 megawatts of wind; by the end of the next year it had 70 megawatts; and by 1985 it had 1,235 megawatts, vastly more than Texas or anyplace else on the planet.

The driving force behind California's wind rush was generous incentives, which no other state came near to matching. In 1978 California lawmakers introduced a tax credit for wind that, when coupled with the federal one, would allow some large investors to knock their total outlay for a big wind project down by 50 percent. Some said it was even more lucrative. One wind hand told Janet Sawin that wealthy investors could get 95 percent of their money back from wind projects even before they began making electricity. By the end of 1983, Sawin writes, the state had made available $100 million in tax credits for wind-power investors. The credits got extended to 1986, but even more crucial was the painfully named Interim Standard Offer Contracts 4 (ISOC4), which took effect in 1983. These guaranteed fixed payments for ten years for wind producers, thus assuring them of a stable source of income during crucial early years, which in turn helped to persuade banks to lend money to the newfangled technology.

Some windmen began trickling west from Texas to join the California rush. One was Bob King, a native of Florida, who had come to Austin for graduate school and had founded the Texas Solar Energy Society in 1977, the same year he wrote a 248-page report called "Alternatives to the Energy Crisis" for the Governor's Energy Advisory Council. King had planned to stay in Texas, and he had a job lined up with John Hill, the attorney general, who defeated Gov. Dolph Briscoe in the 1978 primary and became the Democratic nominee for governor. Hill, with his East Texas drawl, pronounced solar energy—which back then was the term that meant all types of renewables—as solAR energy, says King. "In fact," says King, "I was going to be [Hill's] solAR advisor until he lost by four-tenths of a percent." The victor in the 1978 Texas gubernatorial race, a colorful Republican oilman named Bill Clements, had little interest in alternatives. "His attitude was, 'don't kid ourselves' that there is some magic bullet that will take the place of oil and gas," his energy advisor, Ed Vetter, recalled to Clements' biographer Carolyn Barta.

So King first headed for Knoxville to spend a few years at the Tennessee Valley Authority, which had gotten involved in solar and wind prospecting, and then he pushed onward to the Golden State, where he became a solar advisor to Jerry Brown. In 1982 King founded the California Wind Energy Association. He would return to Texas after Jerry Brown tried and failed to reach the presidency, and more than a decade later King would be a key point man on the state's first big wind farm.

The Carters of Burkburnett also found themselves spending much of their time in California, despite their original theory that their main market would be Texas farmers and ranchers. They soon had turbines up, spinning swiftly, at the California wind farms of San Gorgonio and Tehachapi. These were boom years for the 25,000-square-foot Burkburnett warehouse. According to Carter Jr., in 1981 one of their machines survived 100-mile-per-hour winds, a freak Montana storm that took down trees in valleys. It was a demonstration contest, and other turbines set up in Montana for the demonstration stopped producing electricity. "We walked on water after that," says Carter Jr. "Everyone wanted our machines." Energy historian Robert Righter, however, has a very different account of what may be the same Montana event, drawn from a Montana report and other sources. The Carter machines, he writes, "soon ran into generator breakdowns and the cracking of fiberglass blades. A final, dramatic event occurred on a cold, blustery winter day. A hydraulic-brake mechanism failed and one of the machines destroyed itself, the blades spiraling over the landscape. Carter Enterprises tried to repair or replace the machines, but when failure followed failure they left the field."

Carter Jr. disputes that account—"We were the only one that continued to run and operate in high winds," he insists—and the truth may be lost to the winds. Regardless, Carter Jr. says that by 1982 the company had 800 wind turbines on order, with a backlog of more than two years. Production in Burkburnett reached its peak in 1983, when more than 100 people worked to build as much as a turbine a day, sold as far afield as Hawaii and the Arctic Circle. A few of the machines made their way to Texas farmers and ranchers, but most were sold to California.

        The California boom ended around 1986, after Reagan and Congress allowed a key federal investment tax credit to expire and California's incentives also began to fade. Modest growth continued in fits and starts, but California's total wind capacity actually dipped for several years in the mid-1990s, after the ten-year fixed-price contracts that the state had approved in 1983 began wrapping up, causing wind developers to feel it was not economical to keep their machines going. The Carter assembly line, heavily reliant on California, shut down in 1988. By that point, a squabble had left father and son somewhat at odds. More than two decades later, Jay Carter Sr., eighty-eight years old in 2011, was in ill health but could sometimes be found in the old, cob-webbed warehouse. Carter Sr., who insisted on calling a reporter one-third his age "sir," says their engineering differences were quite small; financing challenges, including the end of the tax credit and a fire at their plant, were the bigger problems that led to them "shutting the business down."

California's wind boom, in the end, served as something of a cautionary tale. The turbines erected to much fanfare during the 1980s had significant problems. In places like Palm Springs, people complained that they were noisy and harmed the desert ecology, but the most fundamental problem was that they often didn't work. They "generated more tax credits than electricity," wrote Matthew L. Wald in the *New York Times* in 1992. In Tehachapi just one of ten turbines installed toward the end of 1981 by a company called Pacific Wind and Solar produced power. Yet all ten—an investment totaling $1 million—qualified for maximum state and federal tax credits, according to Peter Asmus, writing in *Reaping the Wind*.

The lesson was clear: fat investment tax credits, which got doled out based on the amount of money invested and did not require the turbine actually to produce electricity, were not a good idea. They might help rich people slash their tax bills, but that was about it. "Wind machines that never were," Vaughn Nelson called them in a presentation as early as 1982. (Federal wind-power credits, coming a decade later, were oriented toward actual production of power, as opposed to simply planting turbines in the ground.)

"A lot of the wind schemes, and that's what they are, out in California, are simply tax dodges," Russel Smith, a one-time organic fertilizer expert and piano salesman who became a cofounder of the Texas Renewable Energy Industries Association, told the *Texas Observer* in the 1980s. "They are simply money-making things based on the federal government's involvement and the state's involvement in providing tax credits. And some of the stories will stand your hair on end. And also in the solar market. Wind is not exclusively responsible for the abuse."

The California windmills also killed vast numbers of birds, though they did not change the weather by slowing the winds, as some had actually feared. Altamont Pass, in particular, lay in a major migratory path and acquired a reputation as a bird graveyard that it has yet to shake. A study begun in 1989 estimated that turbines there killed 160 to 400 birds each year, including as many as 40 golden eagles. A one-time Sierra Club activist, according to wind chronicler Paul Gipe, took to calling them "Cuisinarts of the air," and the image stuck. The turbines had been erected largely for environmental reasons, and now environmentalists were turning against them. The bird debacle is among the reasons why California has been loath to build wind farms; by the end of 2011 it had barely tripled the wind capacity that existed in the state a quarter-century earlier.

Texas windmen who skipped the California rush were not idle. The 1970s' government-led enthusiasm had dissipated. For example, the successor to the Governor's Energy Advisory Council, which had pumped out reports on renewables, was officially disbanded in 1983. But they were believers, and they still held out hope that Texas government would someday look, again, beyond oil and gas.

The first priority was lobbying. California had fantastic incentives to encourage renewable energy, the installation of it, anyway, even if the machines had flaws. But Texas, which had far better winds, had little.

The 1981 State Legislature had passed useful property and franchise tax exemptions for the purchase of wind-power equipment. And the winds of Texas were getting measured and mapped, thanks to money from an entity called the Texas Energy Development Fund, which lasted from 1977 to 1983. The fund also channeled research money toward the use of wind for large-scale irrigation and, this being Texas, for pumping oil from marginal wells, recollects Milton Holloway, who administered it.

But there was no state tax credit for producing renewable energy, and there were no guaranteed ten-year payments for wind producers. Even the federal incentive arguably most fundamental to the wind industry, the 1978 federal requirement that utilities pay independent producers for their power at a minimum cost, was just that in Texas—the minimum—whereas California had crafted policies that, for a while at least, promised far more payback for putting up turbines. So wind farm developers stayed away: there was "no economic reason for them ever to come into Texas," says Nelson. "[There weren't] any incentives. So they were all going to California."

In 1984, with a Democrat, Mark White, in the governor's office, Michael Osborne, Russel Smith, and another man founded a group in Austin that aimed to shake the legislature into action. Osborne became the treasurer, and Amory Lovins, a Colorado man who would become the country's leading voice on energy efficiency, made an appearance as the Texas Renewable Energy Industries Association was being set up. "Soft energy pioneer Amory Lovins dramatically proclaimed the meeting was one that 'future historians will note as the beginning of the second era of energy in Texas,'" the left-leaning *Texas Observer* reported. The Texans, as usual, were a bit more colorful: "A lot of people think renewable energy is a bunch of balloon-flying hippies humming folk songs," Jim Hightower, then the state's agriculture commissioner and something of a hippie himself, told the *Observer*. "Well, the hippies have gotten down to business."

TREIA supplemented the Texas Solar Energy Society. Most obviously, it had a more apt name. The word "renewables" had gotten around to this industry, which was still dithering over what to call itself, and "once we saw it, we knew that's what we wanted," says Osborne. "It had 'new' in it. You didn't need a focus group on that one. . . . How can you be against something that's renewable? I mean, how can you oppose it?" So enamored was he of the term that he acquired bumper stickers that said "Stop Solar," and would slap them on the backs of Cadillacs just for fun. In 1986 Osborne was written up as a "small-town genius" by the *Dallas Morning News*, which extolled his Pampa wind farm and described some of his solar ventures as "sort of a Bubba meets Whole Earth catalog."

Yet by the mid-1980s the renewables industry was withering. The *Dallas Morning News* correspondent discovered Osborne a little late; by 1986 he had closed the Texaco station, and now his solar dish served as a roof for his living room, turned upside down, like a dome—"'Jerusalem Deco architecture,' he calls it,'" the paper reported. Osborne was still working on solar and had big dreams of a renewable-energy pavilion for the state fair and solar-powered bus stops, whatever that meant.

But in truth, he was headed for a temporary hiatus from building renewable-power projects. His next venture, in 1987, was taking a lease on the historic Austin food and beer hall Scholz Garten. He was joined by Eddie Wilson, of the long-defunct Armadillo World Headquarters, as well as Phil Vitek, founder of the Texas Chili Parlor. Osborne built a stage and added historical refurbishments, but he also wanted to use the place to hold meetings for the new renewables lobbying group. "It was just a nice place to have home court," he says. The venture would tide him over for a few more years, until the oilman Bill Clements (then in his second, noncontiguous, term and enmeshed in a college foot-ball slush-fund scandal) decided not to run again and a Democrat who shared some of his dreams—and often held staff meetings at Scholz Garten herself—took the governorship.

CHAPTER 5

ANN RICHARDS— AND A BIG WIND FARM AT LAST

To keep spirits up at the start of a new decade, Michael Osborne decided his buddies in the renewables game needed a pep talk. Despite the promise of the early work in the late 1970s and the 1980s, no real commercial project had yet been developed in Texas. President Reagan had terminated a big federal tax credit for renewable energy, and the wind industry had drifted into the doldrums. "This decade is going to be good for our industry because we are no longer a bunch of rookies with no experience in the energy business," Osborne wrote in the fall 1990 newsletter of the Texas Renewable Energy Industries Association. "You survivors have the savvy and capability to make things happen." If it was a bullish forecast, it was also a prescient one. By the end of the decade the wind industry would, indeed, win a critical legislative boost.

Still, the industry needed charismatic, visionary leaders like Osborne to keep it engaged. The remarkable thing about the faithful in the renewables industry, like their cousins the Texas environmentalists, is their Sisyphean nature. The culture and history of modern Texas are against them: where they seek conservation and efficiency, they come up against consumption and extraction. This is the land of J. R. Ewing, of exploitation of natural resources and indifference or antipathy to federal environmental regulation, a land of big steaks and heavy appetites. But in 1991, when a steely, steel-tongued politician blazed into the governor's mansion, her administration, at least rhetorically, embraced the notion of clean power. The politician, of course, was Ann Richards, the last Democratic governor of Texas.

Richards grew up in a small town near Waco. After moving to Austin she worked as a junior high teacher before getting involved in local politics, first directing campaigns and then as a commissioner of Travis County, which included Austin. Her husband, David Richards, was a union lawyer, and daughter Cecile Richards once told *Texas Monthly* magazine that her first dance, held when she was just nine, took place at a VFW hall in the South Texas town of Mission, where Chicano farmworkers were preparing a protest march.

Ann Richards had her struggles—a 1984 divorce, alcoholism—but after winning office as state treasurer, she burst onto the national scene in a cutting 1988 speech at the Democratic National Convention. "After listening to George [H. W.] Bush all these years I figured you needed to know what a real Texas accent sounds like," said Richards, who with that speech became known around the country as that Texas woman with the frosted hair. Bush, she famously added, "was born with a silver foot in his mouth."

After Democrat Michael Dukakis lost to Bush in the 1988 presidential election, Richards was suddenly a source of cheer for a down-and-out party, perhaps even vice presidential material if she could keep it up. "Leaderless at the national level, backpedaling in the face of George Bush despite controlling both branches of Congress, and clueless as to what to do about losing five of six presidential elections, the Democratic Party is like a brontosaurus," a *Boston Globe* columnist wrote in a glowing profile of her in 1989. "There's the huge body, the vast appetite, and the tiny pea-size cranium controlling that monstrous tail." The following year she won the governorship after a bruising primary and general election in which her Republican opponent, Clayton Williams, called her a liar and refused to shake her hand following a joint public appearance. He was well on his way to losing the women's vote when he compared rape to bad weather. "If it's inevitable," he had said, "just relax and enjoy it."

Richards's victory was narrow, but when she took office, she shied not a whit away from liberal talking points. Among these was her stance on environmental issues. Her 1991 State of the State Address promised to "build a safe environment for Texas in harmony with a strong economy." "No longer will Texas be known as a toothless tiger when it comes to penalties for polluting," she thundered. "No longer will Texas be held hostage to polluters—often out-of-staters, their carpet bags stuffed with hazardous waste—taking advantage of lax environmental regulation." Appointees would be environmentalists, she said during that address, and they would be held accountable for their performance.

In July of 1991, only six months into office, Ann Richards's staff announced a pilot project to install three wind turbines, designed and manufactured by International Wind Systems of Burkburnett—headed by Jay Carter Jr.—northeast of Amarillo. In retrospect, it was a modest proposal, but one trumpeted with all the heraldry a governor can muster. "Texas must develop a cleaner, more diversified energy base," declared Richards. "Our state is blessed with an abundance of renewable energy resources—like the clean, inexhaustible Panhandle winds—and my Energy Office is vigorously pursuing their development. Projects like this one protect the environment, benefit consumers, and demonstrate the emergence of a cutting-edge renewable energy industry in Texas."

Behind the proenvironment hurrahs, it was a crisis of psyche in big-hair, oil-rich Texas that led some officials to more serious consideration of wind power. By the early 1990s Texas, the energy state, had become a net importer of energy due to declining production and the continued hunger for fuel to supply the refineries and chemical factories clustered along the coast. It consumed more energy than any other state in the country, yet in 1992 oil and gas production accounted for only 11 percent of gross state product, down from a high of 26 percent in 1981. "The United States, which once exported Texas crude oil, now exports Texas oilmen," Matthew L. Wald wrote in the *New York Times* in 1992. "We thought it a shame—and a motivator," says Karl Rábago, a former military man who had taught law at West Point and whom Richards appointed to the Public Utility Commission of Texas, the agency responsible for overseeing the electric sector.

In addition to this embarrassing state of affairs, another force beyond Texas began shaping national and state energy policy in the early 1990s. Over the previous decade, scientific consensus had formed that humans were playing a role in the warming of the planet. Just as Richards took office in Texas, nations across the globe were preparing for the Earth Summit, a 1992 conference in Rio de Janeiro. So major was the event's promise that the *New York Times* alone had dozens of stories in the lead-up to and coverage of the climate talks in Rio. Much of the coverage raised the profile of renewable energy as an alternative to fossil fuels. A study by the Union of Concerned Scientists and a handful of other environmental groups argued that if the United States promoted energy efficiency and renewable energy, the country would end up with net savings of $2.3 trillion over forty years and 70 percent fewer carbon-dioxide emissions. "To ignore the economic opportunities is to fail to seize the moment, to become paralyzed by exclusive focus on one side of the economic ledger," Harvard economist Robert Stavins told a conference on global warming in Washington.

The United States, with its heavy manufacturing and refinery base in Pres. George H. W. Bush's state of Texas, was not entirely sympathetic to such reports. By the time Bush landed in Rio in June to address the conference, the United States was widely condemned as a villain as it negotiated for a looser greenhouse gas treaty.

The United States had shown interest in cleaning up the air. Two years before Rio, Bush, with backing from huge majorities in Congress, had signed a landmark update to the Clean Air Act, which established a national emissions-trading program designed to bring down the levels of chemicals like sulfur dioxide and nitrogen oxide that cause acid rain. Global warming, however, was a trickier prospect, involving atmospheric gases whose effects Americans couldn't immediately see.

Ultimately the United States did sign Rio's climate treaty, along with 153 other nations. Under the treaty, a predecessor to the 1997 Kyoto Protocol, nations pledged to devote money and technology to stabilizing the world's greenhouse gas emissions in the hopes of avoiding artificial disruption of the climate. But the treaty set no mandatory limits, and to this day carbon dioxide remains largely unregulated in the United States.

Texas renewables advocates added the global warming argument to their cause and knew they had a sympathetic ear in a governor known for her love of camping. As far back as the April 1986 TREIA newsletter, Osborne, reviewing a federal report on climate change, wrote, "So, great! We do away with coal and add all of our new electric capacity by erecting wind parks in the wind regimes and solar parks in the solar regimes." Now, finally, the renewables industry had real international attention and a governor who seemed interested in environmental and energy policy. "The good old boys had brought us lignite and overpriced, poorly run nukes," says Rábago. Now, finally, there seemed to be a chance for a new energy policy.

And so Richards's administration decided to push for renewable-energy production. Her timing was in sync with the feds. In 1992, as part of the Energy Policy Act, Congress passed a national renewable-energy production tax credit, which would guarantee a payment of 1.5 cents per kilowatt-hour of energy produced by wind farms for the first ten years of their existence. This would not be a repeat of California, where an investment tax credit threw masses of money at wind projects that often didn't work. This would be a production incentive, more effective than the earlier investment incentives. To get it, the farms actually had to generate electricity, and the credit would not expire until 1999. (In fact, the credit has been renewed multiple times since then, though often haltingly and sometimes after the expiration date, to the immense frustration of wind developers and manufacturers, who view it as crucial for a still-youthful industry.)

Richards's first act was to pull great minds together to figure out what should be done. Early in her tenure she established the State of Texas Energy Policy Partnership, or STEPP, to develop a broad energy policy for the state. The partnership was meant to have only 25 or so members, remembers Carol Tombari, who headed Richards's energy office, but "Governor Richards being a politician, it ended up with 300 members." Plenty of old faces reappeared: Vaughn Nelson chaired one subcommittee, Michael Osborne, another. (Osborne had known Richards since the Armadillo days and, perhaps mindful of her political career and her drinking issues, he had persuaded her against entering into the Scholz Garten partnership a few years earlier.)

"The only real efforts that I've seen at a national energy policy amount to a declaration of war in one place or another," Richards told the convened members of the partnership on June 16, 1992. "What we do to get cheap energy is to send naval warships over to the Middle East to escort out the oil tankers . . . and bring that oil in here to compete with domestic energy. If we consider the price that we pay for those naval warships in their escort capacity, this state would never have had a recession."

The STEPP process delivered mighty reports to the governor's desk, urging more action to make the state a technology hub for all forms of energy. But Richards wanted more work on renewables in particular, and in 1993 she signed an executive order creating the Texas Sustainable Energy Development Council, charged with crafting "a strategic plan to ensure the optimum utilization of Texas' renewable and efficiency resource base"—in other words, figuring out how Texas could move toward renewables. Rábago co-chaired it along with Garry Mauro, the head of the General Land Office, a state agency in charge of caretaking, leasing out, and gathering royalties off of state-owned lands. Mauro, who had earned high school football stardom in his native Bryan, was a staunch Democrat and a lawyer and had been elected Texas land commissioner when he was just thirty-four.

The Sustainable Energy Development Council produced, among other things, a thorough report on the renewable-energy resources in Texas. Mike Sloan, a young renewables expert with a pair of degrees in mechanical engineering, crafted the report, and it found—much like the Nelson/Gilmore report had two decades earlier—that Texas possessed "vast areas with high wind power potential." But Tom "Smitty" Smith, an environmentalist with Public Citizen, was chagrinned that the report did not get released until after Richards's tenure was up. With George W. Bush, the folksy son of the man born with the silver foot in his mouth, closing in on her, "somebody decided in her campaign that it would be too controversial," Smith says. "They got printed in the interim between the time she lost the election, and when Bush took office."

In some respects, then, the Richards administration did little for renewable energy. While turbines continued to spring up in California, albeit more slowly, and Minnesota developers built a 25-megawatt wind farm in 1994 in a spot called Buffalo Ridge, the high-water mark for Texas during those years amounted to a trio of Carter turbines near Amarillo operated by the Southwestern Public Service utility, another tiny cluster of Carters erected by Texas Utilities near the Dallas–Fort Worth airport, and a couple of bound reports. "I had frustration that as the director of her energy office I never could get her attention," said Tombari, who now works on energy projects in Colorado. "If I had a court filing to deal with, I could never get Ann's legal counsel, because he had a midnight execution he was trying to stay. Quite frankly, Ann Richards had other things to deal with."

But the Richards administration also managed to blow vitality into the renewable-energy industry and the wider environmental movement. "Ann Richards was the one that put renewables at the table," says Michael Osborne. Emerging from the trough of the 1980s, when federal and state policy makers alike had shown little interest in changing habits or energy resources, the Richards officials and their stack of reports propped up the movement in a way that galvanized those interests for battles farther down the road. The State of Texas Energy Policy Partnership, for example, concluded that Texas ought to think of itself as an energy state, so to speak, not just an oil and gas one. It was a small distinction, but it helped entrench the no-silver-bullet-argument that is heard around the capitol even today: that the state needs renewables, along with coal and oil and gas and nukes, to meet the needs of Texans. At the very least, Richards's administration set the stage for the first of several breakthroughs that would set Texas wind on course to national dominance. Six months after she left office, a Texas utility would erect the largest wind farm outside of California in a grand, sometimes terrifying, experiment.

One of the people who would put this experiment in place was Dale Osborn, the president of a California company called Kenetech that had become the nation's largest wind turbine manufacturer, with a factory in the Bay Area town of Livermore. Kenetech, formerly U.S. Windpower, operated with something of a locker-room swagger. "This is not a fraternity, this is a highly competitive business arena—and we are now the most competitive organization on this globe to deal with wind issues," Osborn had said in one interview. In a swipe at former California governor Jerry Brown, who had advocated early subsidies for renewable energy, he declared that the wind industry "can't afford to look like Governor Moonbeam's children. This is not a ponytail industry."

U.S. Windpower had been founded in Massachusetts shortly after the oil embargo by engineers Stanley Charren and Russell Wolfe, who dreamed, like so many others, of a country that consumed far less fossil fuel. But unlike so many others they had found success: U.S. Windpower had installed the twenty wind turbines, rated at 30 kilowatts each and perched atop sixty-foot towers, on a slope of Crotched Mountain, New Hampshire, in late 1980 and early 1981—creating a "'farm' that has nothing to do with raising vegetables, fruit, grain, or livestock," as a *Christian Science Monitor* article at the time helpfully explained. It was the first wind farm in the country, beating Michael Osborne's smaller project in Texas by less than a year.

True, the turbines had fared badly and the venture ended not long after it began, but the company, unfazed, had continued its expansion in California, where the soils for a wind company were rich with incentives, and where it began improving its blade system and selling new models of turbines. Some U.S. Windpower machines were among the earliest to go up at the Altamont Pass wind farm. In the late 1980s the company reorganized as Kenetech as it worked to secure future sales contracts and steamed ahead with research and development; by 1991 Kenetech had achieved a profit of close to $10 million.

The R&D appeared to pay off in 1993, when the company proudly rolled out its new, revolutionary model, the 33M-VS, with a 33-meter rotor diameter and variable-speed drive technology. Until the 33M-VS, wind turbines moved at a fixed speed and could not go faster in response to changing winds. "When the winds speed up, the turbines don't speed up, losing that extra speed and energy," Osborn says. A key innovation of the new model was to convert variable frequencies to a fixed voltage, thus enabling extra power to be captured. Without a steady voltage, surges and sags would have hit the transmission lines that moved the power from the turbine. "Think of it as a big water hose with constant pressure," says Osborn.

Because it was more efficient, the new turbine could generate electricity at 5 cents per kilowatt-hour, Kenetech officials said, instead of the 7.5 cents per kilowatt-hour or more its previous machine demanded. With the board thinking it had hit a home run, says Osborn, the company did not put in the hours to examine whether there were any manufacturing or design problems, an oversight that would lead to problems later.

Fueled by bravado, the hope of a resurgent California, and the need for more money to fund mass production of the new model, the company had a wildly successful public offering in 1993, selling roughly six million shares for as much as $25.50 per share. Merrill Lynch foretold a hundredfold rise in Kenetech's sales over three years, according to journalist Peter Asmus, who has a tidy recounting of the Kenetech saga in *Reaping the Wind*.

But Osborn, a Vietnam vet, felt that the company had mentally hemmed itself into California, so in 1992, he rented an apartment and office in Austin with the aim of expanding operations. It felt like a kind of renewable-energy homecoming. Osborn had first grown curious about wind turbines while working in Lubbock for Texas Instruments in the early 1980s. A couple of his kids went to Catholic school, and it was there that he first laid eyes on the Carter turbines erected by Father Joe James on the grounds of the quirky church, Saint John Neumann. Now with Ann Richards in the governor's mansion and Texas groping to answer the existential question about whether it would forever be an importer of energy—a question that drove at the very pride of the state, as well as its economics—he decided Texas was ripe for development. The news that Texas was a net importer "had caused some tremors, and people were thinking, 'Maybe our economy shouldn't be totally fossil-fuel based,'" Osborn says. First, he recalls only half-jokingly, he had to convince his own board that there were opportunities outside the Golden State. "Our board did not think there was a potential wind industry outside of California," Osborn says. "Nobody realized wind blew in other places outside of California."

But Osborn lacked a partner. In the early 1990s he pitched the idea of a commercial wind farm to a half dozen utilities around the country, and none wanted to take a risk. A utility in North Dakota "thought I was crazy," he says. "They said, 'We'll do everything short of murder to keep you out of our market.' If crazy Californians were doing that, then anything associated with you was nuts. These people were ground in, from coal country in North Dakota." He had similar experiences with utilities in other states, including in Texas. "Utilities are control freaks, fundamentally. They have to be that way by necessity, to keep the lights on. Their job is not to be all that creative or innovative, to make sure prices are low, to make sure the system is highly reliable. Their motto is, 'Let's not get in trouble.'"

Then, in Austin, as Osborn schmoozed about in an effort to make contacts, he met Bob King. King helmed Good Company Associates, an Austin consulting firm specializing in renewable energy and efficiency. He had grown up in Cocoa Beach, Florida, within twenty-five miles of the space pad that would send men to the moon, which meant that he had rocket scientists for neighbors and was building solar-powered gadgets by the time he was in the seventh grade. King had gotten involved in renewables in Texas before wandering away to the more promising markets of California. After he came back, he spent several years working for a lefty Texas agricultural commissioner named Jim Hightower before founding Good Company, which helped create recycling programs for big names like Frito-Lay and Levi Strauss, and, as an internally progressive measure, rewarded employees who did not drive to work.

"The name came to me one morning, I was meditating. . . . I had been trying to think of what I would call it," said King, in the kind of vague meanderings of smarts, hippiedom, and do-gooding that epitomize Austin. "I wanted to do good things, I wanted to prove you could do good things and make good money. And I really wanted to have a company that was a good company and treated people right. And I wasn't exactly sure what I'd be doing but this was broad enough to cover it."

That way of thinking was good enough for Osborn, who hired King in the early 1990s to put together a deal for Kenetech in Texas. The challenges were to find a utility to partner with Kenetech and to figure out a way to get the power from West Texas to the population centers in the central part of the state. For King, working for Kenetech, a company that dominated the youthful wind industry, felt cutting-edge. "Kenetech was like Enron, sort of—not in a negative way," King says. "Everybody was buzzing about what Kenetech was up to."

Perhaps the most crucial work King did for Osborn was to introduce him to one of the great Austin political fixers, Ed Wendler Sr. Wendler, the son of two reporters, liked to call himself the King of the Lobbyists. *Texas Monthly*, in a story about growth in suburban Austin, described how one area came to be called "Wendlerville," thanks to his close involvement in certain developments. A 1987 *Austin Business Journal* article described him as a distinctively Austin creature: "stridently liberal in his political thinking and, in the same breath, the lead cheerleader in the city's move toward development, growth and land flipping that turned baby-faced builders into instant celebrities." He demonstrated for Democratic rights for fledgling Latin American democracies, worked to register voters in parts of town where residents had once been denied rights because of their race, and helped devise a "gentleman's agreement" in Austin politics, which held that one spot on the city council went to a Hispanic and another to an African American. "The nickname they had for him was 'the shadow mayor,'" Bob Mann, an Austin political consultant and University of Texas journalism teacher, told the *Austin American-Statesman* after Wendler died in 2004. Wendler enjoyed an easy proximity to power. A friend of Lt. Gov. Bob Bullock, the blunt, strong-willed legislative leader, and an occasional presence at the Armadillo during the 1970s, Wendler once said, "You can't be a bleeding heart with no money."

Wendler, so went the plan, would help Kenetech find a partner to build its wind farm. After all, he knew just about everyone in town. Dale Osborn says that while he was having dinner one night at the Austin music joint La Zona Rosa with Wendler, Ann Richards stopped by their table to say hello to Wendler as she was walking out. "Governor Richards, I'm really happy to have had a chance to meet with you," Osborn remembers telling her. "And she said: 'It's always nice to be met.' She was such a gracious lady, she didn't know me from Adam." The story illustrates the sort of entrée Ed Wendler could win a client.

But wind was new and risky, and Wendler would have to find somebody who would promise to buy the power before a developer invested millions in building a wind farm. And this person needed to be free of the shareholder pressure and narrow thinking characteristic of most utilities.

      At a utility headquartered beside a lake in Austin, Wendler found the perfect candidate. The Lower Colorado River Authority (LCRA) is one of the most political institutions in a highly political town. Its storied history begins in the 1930s, when the state decided to build massive dams that would hold water, control the flooding that plagued places like Austin, and deliver hydroelectric power to the impoverished Texas Hill Country—a cause championed by an aspiring young congressman, Lyndon B. Johnson.

The LCRA supplies water to Austin and to rice farmers near the Gulf of Mexico, and it also feeds electricity into the Texas power grid, not just from dams, but from coal plants and gas plants, too. Though established by the state and overseen by a board appointed by the governor, it is officially a nonprofit and relies on the sale of water and wholesale electricity, not tax dollars, to pay for its operations. Essentially, the LCRA is a miniature, state-level version of the Tennessee Valley Authority (TVA), the huge federally owned utility that supplies power to much of the South.

The ability to generate electricity and dole out water has endowed the LCRA with enormous political power, but it also needed to be on the receiving end of a few favors to get things started. To secure millions in federal help during the Depression, for example, early LCRA organizers offered to name a dam after Texas congressman James Buchanan, who happened to chair the U.S. House Appropriations Committee, key to approving Pres. Franklin D. Roosevelt's New Deal programs. In June 1936 Buchanan met with Roosevelt and gave the following account, as recorded in the first volume of Robert Caro's commanding biography of LBJ: "'Mr. President, I want a birthday present.' 'What do you want, Buck?' Roosevelt is said to have replied. 'My dam,' Buchanan is said to have answered. 'Well then, I guess we'd better give it to you, Buck,' the President is said to have replied; picking up the telephone, he gave Secretary of the Interior Harold Ickes the necessary order." When Buchanan died of a heart attack in 1937, Lyndon Johnson, then the Texas director of a Roosevelt agency that put young people to work, ran for—and won—his seat.

But by the mid-1980s, the political audacity that had characterized the LCRA's formative days of early expansion had slid into boorish back-slapping. In 1986 a five-month investigation into LCRA activities dating to 1981 found a conflict of interest and favoritism with the awarding of a major contract, sexual improprieties, and mismanagement. The scandal was collectively known as "trailer-gate," after a series of episodes involving booze, gambling, and sex in trailers belonging to LCRA contractors.

"There will be a new team of leaders on my staff, including existing employees," new general manager S. David Freeman, an experienced energy hand who had chaired the board of the Tennessee Valley Authority and was brought in to clean up the LCRA mess, told the *Austin American-Statesman* in 1986. "They will be people who know right from wrong in a public agency." Freeman, who had been recruited to the job by fellow TVA veteran Bob King, quickly fired some LCRA officials as he worked to restore the confidence of the public and the legislature in the group. At one point, faced with an especially dubious state senator, "I took all the American Express cards of the employees, collected them all . . . cut them all in two, put them in a brown bag and handed it to the senator and says we've cleaned up our act," Freeman recalled two decades later. "And he took that bag of credit cards all over the state."

At the same time, Freeman began recruiting people he thought would return the utility to its former glory. Among them was Mark Rose, who had promoted energy efficiency as a popular Austin City Council member in the mid-1980s. Rose, sandy-haired and boyish-looking, joined LCRA in 1987 as Freeman's deputy. He became his successor in 1990.

Rose was a history major who had a keen sense of the obligations of public power. "We were founded on renewables," says Rose. "We consider water to be a renewable. Economic development was part of our mandate." He had been going out to Far West Texas since he was a teenager on jaunts with his mother, an ardent bird-watcher and artist. His pedigree also prepared him to think comprehensively about wind power. His father, a brigadier general named James M. Rose, had headed an energy conservation task force under Gov. Dolph Briscoe during the 1970s energy crisis. Like Dale Osborn, Rose had come to view utilities as "reactive" and risk averse, and if he could take a step out of that mold in the public interest, all the better.

Wendler made the connection, bringing Osborn and Rose together. To get a Texas utility like the LCRA to invest in a wind farm, the land would have to be within the state's boundaries. Wendler, handy again, brought in his old law partner, Garry Mauro, who happened to be land commissioner and in that position was responsible for raising money for the schoolchildren of Texas through the prudent management of the state's millions of acres of property.

During dinners of smoked meat at the Iron Works, a red tin shack of a barbecue joint named for the blacksmithing work of a German family, Osborn, King, Wendler, Rose, and Mauro crafted strategy. Kenetech had a manufacturing facility in Livermore, California, that could roll out the turbines. With the new turbine technology, a utility using Kenetech machines could sell wind power at roughly 5 cents a kilowatt-hour, an amount low enough to be palatable to the LCRA, a wholesale electric provider. Mauro would offer the imprimatur of the General Land Office on the land in Far West Texas that the project would lease. This mattered because the power had to be transmitted, somehow, hundreds of miles east, and there was no guarantee that the local utility, which owned the power lines near the site, would allow it across (a concept called "wheeling" the power from one jurisdiction to another). If the state was involved, it could essentially force the utility's cooperation, because King and Wendler, with backing from Mauro, had sneaked a provision into a 1993 law that legalized this wheeling as long as the power originated on state lands. "It got buried in there and nobody knew what the hell we were talking about," says King. "Nobody was really threatened by wind power at that point."

But there was another problem: the location they eventually settled on wasn't technically state land. State land would have been ideal, King says, but the holdings controlled by the General Land Office in West Texas were somewhat fragmented because of the way land had been parceled out to the railroads more than a century earlier, and the state didn't have a large enough plot that also had the right winds.

So a complicated deal ensued. The landowner agreed to lease the land to the state, which then leased it to Kenetech to build the wind farm. A new electric transmission line was to be built, quickly, to move power to the grid. Those involved still snicker at the memory: it wasn't entirely aboveboard because of the way the transmission line was privately financed and later sold to the LCRA, but it worked—and the LCRA gamely endured the expected tongue-lashing after the fact from the state's electric regulators. ("We caught the Eagle Scout smoking a grapevine behind the scout shack," one state regulator told Rose, according to Robert Cullick, another LCRA employee.)

One Saturday morning in Mauro's office in 1993, over a pile of barbecued chicken, as Osborn remembers, the group settled on the deal. First, though, the LCRA would put the project out for public bid. "Not since LCRA's hydroelectric plants were installed on Central Texas' Highland Lakes dams several decades ago has there been electric generation with renewables on a comparable scale," Michael Osborne wrote in the summer 1993 TREIA newsletter.

Not surprisingly, the favored bidder was Kenetech, and that November Kenetech won the right to build the 35-megawatt wind farm. At a cost of about $40 million, 112 turbines would be erected in the Delaware Mountains on a wind farm called the Texas Wind Power Project, feeding power into a twenty-eight-mile, 138,000-volt transmission line that cost $6 million and would hook into lines owned by Texas Utilities. Once it hit the main grid, the power would shoot east all those miles to the LCRA's territory in the heart of Central Texas (in theory, at least; in practice, the electrons cannot be quite so easily controlled). The state would get royalties from leasing the land, as much as $3 million over twenty-five years, which would go to the state's Permanent School Fund, which helped public education. It was, Mauro said, a "whole new era" for a fund that had built itself up with oil and gas revenue.

Kenetech had wanted to build something even bigger—50 megawatts, the company had said, was the minimum it had to sell to make the project economically viable. The LCRA, eager to move forward, agreed to buy 25 megawatts of the power and recruit other takers, so Rose and Tom Foreman, then the LCRA's manager of resource planning, visited the chief executives of major utilities, including Houston Power and Light and Texas Utilities, to try to get them to join. "Mark was selling them—'Hey we think this would be a good thing for the power industry,'" Foreman recalls. But the utility executives, he says, "kind of thought we were nuts." One buyer did appear: Austin Energy, ever eager to please its green constituency, agreed to take 10 megawatts. So the project was on, with the LCRA buying 25 megawatts and Austin Energy picking up a bit more. It added up to enough power for at least ten thousand homes—not a vast amount in the scheme of things, but a start—and if it worked, the project could be massively expanded.

For Rose, the head of a major utility, the price was still a bit of a stretch, but he had no second thoughts about diving in. With the help of the federal production tax credit of 1.5 cents per kilowatt-hour factored in, the LCRA would pay about 5 cents per kilowatt-hour, roughly double what it paid for its usual mix of coal, hydro, and natural gas, plus another cent to cover the cost of the transmission spur. "I literally said to my board, 'These are spiritual megawatts. Don't ask me to justify them'" on an economic basis, Rose recalls. The fact that Ann Richards was in office while the LCRA was planning its deal gave Rose the political cushion he needed in case the board didn't see things his way. "Mark Rose could feel very comfortable that the LCRA wouldn't oppose him building a wind-power plant, because he would always have the option of calling Ann Richards and saying, 'Ann, your appointees aren't helping me,' or, 'Ann, we need to appoint some new people who will be more supportive,'" says Mauro.

But the board was "very enthusiastic," according to Rose, and it probably helped that the 25 megawatts the LCRA would buy represented just a sliver of its total power supply. "Who the hell is going to notice 25 megawatts at X amount," Rose says. "I mean, it's in the noise. What's important is that this project will capture attention. What's important is that this project has the potential of waking the industry up. And that is well worth the difference."

To get to the Delaware Mountains from Austin you must drive eight hours, across the rolling Hill Country and then, once it smoothes out at the town of Junction, through the vast Chihuahuan Desert on the seemingly unending I-10. You pass through scattered mesas and through the town of Fort Stockton, briefly a Confederate outpost during the Civil War. You are on the frontier now. But you're not there yet. Next you drive through the biblical-sounding burg of Balmorhea, a cold-spring oasis in West Texas. At a place called Van Horn, named for a nineteenth-century commander of the Fort Bliss military base, you take a right to head north on Texas 54. Take that for an hour through a lot of nothing much. And before reaching the New Mexico line you are at the foot of the Delaware Mountains, an old, massive, straight-up coral reef from ancient days when this part of the world was under water. The distance between the Delaware Mountains and Austin is the distance between New York City and Cleveland, Ohio. It was this distance that Mark Rose and Dale Osborn decided to span in 1995 with the state's first commercial wind-energy project.

Osborn negotiated with a sharp septuagenarian named Tony Kunitz to lease the land. Kunitz, who had made his money laying gas pipelines along the Texas Coast, had bought 85,000 acres of the inhospitable land in the 1980s to take his clients deer hunting. The mountains held secrets that were whispered about on the lonely back roads: a B-17 bomber was said to have crashed into them in the 1940s; an old limestone shack had been home to outlaws on the lam; and the cattle were angry to be in such a godforsaken place, jokes Osborn. To get to Kunitz's place, on a mile-high treeless ridge known as the Six Bar Ranch, you had to take a harrowing switchback road of loose shale. Flat tires were almost a given, and that was the least of anyone's worries. "If you've ever driven on that, you know once you start sliding, you can't stop," says Tom Foreman, Rose's right-hand man, still shuddering at the thought of the drive fifteen years later. "There was no guardrail. Oh my gosh."

A bigger road up from the salt flats was needed, one that would be wide enough and sturdy enough to bear the flatbed trucks that would carry turbines to the ridgetops. Kenetech hired an outfit that built roads through the Rockies to blast a path, because the trucks, with their huge loads, "couldn't be going up the tiny switchbacks," Foreman says. Osborn imagined you could see those jags from outer space. "We often say, 'You can build almost anywhere, that's why God made bulldozers,'" jokes Osborn.

The Delaware Mountains site, 100 miles east of El Paso and about 50 miles north of Van Horn, was the windiest spot that could be found in Texas, save for the area around nearby Guadalupe Peak, which was off-limits because it was in a national park. "It was a great site," says King. "I mean, if you got out of the car with the car facing the wrong direction, it'd rip the door off on a good day." Osborn believed it was the "second-best wind site in the United States, second only to Wyoming," as he told the *San Antonio Express-News*. It far outdid California, where the wind speeds at Altamont Pass averaged sixteen miles per hour, versus twenty miles per hour at Kunitz's place.

Michael Osborne had actually found the site a few years before. While driving around for Osborne Solar, in the days when he would slap a makeshift sign that read "U.S. Renewables" on his car while hunting for windy prospects to flag for clients, he found an abandoned gypsum plant in the small settlement of Elcor, about forty miles northeast of Van Horn. It was in some salt flats near the mountains, and it had, most importantly, a power line running out to it. The power line meant that the place, for all its desolateness, was actually hooked into the grid—and a wind farm could potentially feed power into the grid also. Upon spotting the line, "I went, 'OK, that's our railroad—that's our link to those Far West Texas winds,'" Osborne recalls. He got permission to erect a tower to put up wind-measuring equipment: a big box, slightly larger

5.1. *The LCRA wind farm in the Delaware Mountains, in Far West Texas, was the first major project of its kind in the state. Courtesy LCRA Corporate Archives (EU00627).*

than a shoe box, that Vaughn Nelson, the West Texas State professor, had distributed when he was trying to get more wind readings than just the National Weather Service's data at airports. Osborne persuaded a "little lady" in a highway patrol station on Guadalupe Pass, roughly sixty miles to the northwest, to take more measurements there. The readings from the area were incredible—the winds were nearly off the charts—but Osborne says he never made money off finding the place.

Kenetech decided it was the perfect spot for its project. The locals mostly came around—John Conoly, the Culberson County judge, told the *Express-News* that while some ranchers had initially opposed the wind farm, because of tax abatements the county had doled out, the local businesses were for it. And on August 31, 1995, the turbines first began producing power. Mauro, the land commissioner, hailed the event as the "beginning of a sustainable energy resource that is cost-competitive and will keep Texas economically viable after its fossil energy resources have played out." More than two months later, all 112 turbines were operational, making it the country's largest wind farm outside California. The opening ceremony was held on November 14, 1995. So remote was the spot in Culberson County that three counties had to be scoured before enough nice linen could be found for tablecloths.

Not that the tablecloths flapped that day. All the important officials assembled, but "there was zero wind," says Foreman. "So here we were at Texas's first large-scale commercial wind plant, and there was zero wind. Kenetech finally took one of the turbines and motored it [i.e., they set it spinning despite the lack of wind] so you could see what it was like when it was turning. It was beautiful, the sun was out, it was cool, and it was dead calm." From the desert floor below, wrote Bruce Hight, an *Austin American-Statesman* reporter, the eighty-foot-tall turbines appeared like pinwheels atop a fence. "These are like my children," Dale Osborn, who was already dreaming about expanding the project, told Hight during a 1995 visit to the site.

Only a couple of months after the opening ceremony, those children, the turbines Osborn had worked so hard to erect, were nearly obliterated. A monster winter storm blew through West Texas. Gusts reached up to 163 miles per hour in the Delaware Mountains, equivalent to the sustained winds of a Category 5 hurricane. Southwest Airlines halted all flights out of El Paso as tornadoes ripped through the region, and sandstorms kicked up by the same system knocked out car windows and power lines and even toppled trucks.

Phil Davison, a correspondent for the British newspaper *The Independent*, was driving to El Paso from Mexico. "By the time we stopped for coffee and petrol, a mean wind had whipped up and tumbleweed had piled up against the car," he wrote:

> *Most tumbleweed I'd seen was in the Peanuts cartoons, drifting past the Arizona home of Snoopy's brother, Spike. But as we drove north up Highway 45, we ran into the kind of tumbleweed we never want to see again. Or rather, it ran into us.*
>
> *There were hundreds of giant balls of twigs crossing the highway every hundred yards and traveling at about 50 miles an hour. They got bigger and bigger until some were 10 feet high, and not "weed" but bunches of hard scrub rocking the car with each hit.*
>
> *Then came the sandstorm. Within minutes we were down to crawling pace, our emergency lights flashing, peering but barely seeing beyond the bon-*

> *net. When we saw flashing lights, we stopped and*
> *discovered others had collided off the road.*
>   *I'd been through sandstorms during the Gulf*
> *war, but there's a difference between being in a*
> *sturdy tent with American Marines and armoured*
> *vehicles outside, and being stranded in a small*
> *saloon car with sand piling up against your door.*

When the lights flickered back on, officials began to tally the damage. In Anthony, twenty-six miles west of El Paso, a grocery store roof had collapsed, killing a woman and badly injuring at least two others. Telephone poles had snapped "like matches," according to LCRA spokesman Bill McCann. El Paso police set up a "war room" to deal with the chaos.

The day the storm hit, January 17, 1996, Tom Foreman was reporting to the Lower Colorado River Authority's board about the success of the brand-new wind farm. In three full months of operation, he told the board, the project had generated over 30 million kilowatt-hours of electricity. (That's enough to supply about 10,000 modern households over that time period.) "As I came out of the board meeting, a TV [was playing] out in the foyer," Foreman recalls. "Someone said, 'Isn't our wind plant out near El Paso?' I said, 'Yes, why?' They said, 'Eighteen-wheelers are being blown off the road there.'"

Foreman saw on his meters that production from the wind farm had stopped. Frantically he called Kenetech but could not get through. He immediately began worrying about liability, but the LCRA had none— when it made the deal with Kenetech, it agreed only to buy the power, not to take responsibility for the project itself. That move seemed wise at the time, and the reality of hurricane-force winds made it seem wiser still, though Foreman and Rose certainly did not want to see their grand experiment fail in the Texas weather and face a chorus of "I told you sos."

As it turned out, Kenetech was dealing with the most worrisome problem of all. At the time the storm blew through, there were men on the mountain—sometimes called windsmiths—repairing the turbines, and the company, according to Foreman, was urgently trying to locate them. They could not get down the mountain because of the wind and the dust, and had thought to hide in a big metal container that kept tools. That blew away before they climbed in. So they took refuge on the mountain, in a sheltered spot.

Some of the turbines, of course, did not survive. As an LCRA spokesman, Bill McCann, told the *Statesman* at the time, the machines were designed to stop working when the winds hit 65 miles per hour and to be able to last through 120 miles per hour. But "nothing is designed for 163 mile-an-hour winds," McCann explained. At the wind farm, two towers fell, two others got bent and a few dozen others suffered various degrees of damage. Back at LCRA headquarters, Foreman saw the pictures of some of the damaged turbines. They looked, he thought, "like crumpled flowers." Remarkably, within a week or two, the plant was churning out a trickle of electricity again. Other key pieces of equipment, including the substation and power lines, survived unscathed, and the wind farm itself was quickly rebuilt so that by late February, 85 of the 112 turbines were working again. Within months, the plant resumed full operation.

For the wind industry, this was a clear lesson on site placement. "One of the things we learned," King recalls with a chuckle, "was you don't necessarily want the windiest place."

Meanwhile, however, King's employer was sliding rapidly into bankruptcy. King says that by the time of the storm, Kenetech—which had built the turbines as well as developed the wind farm—had already sold the project to investors, even though it was still operating the facility. Its business model was to find markets for the turbines that it made and then offload its ownership of the wind farm. "As quickly as we could put together the three elements of a deal (find a site that was windy, sign up a customer, and contract for delivery), we would find someone to be the owner," King says. "We brought in equity owners who could benefit from the tax credits, and debt for the remainder. We took the money, including the development fees or profit, right up front, and used it to grow and go find the next projects." King's recollection is that the new owner, LG&E, probably "ate the cost" (or at least some of it) of the damage from the storm, as insurance was unlikely to cover off-the-charts wind speeds.

But Kenetech was in trouble in other ways. The company was already reeling from its inability to successfully close deals in Spain and India, and in February 1995, only about six months before the first turbines began spinning in the Delaware Mountains, Kenetech suffered a blow in California, the nation's biggest wind market, when efforts by state energy regulators to compel the state's largest public utilities to buy more power from independent producers got blocked by federal authorities. Pacific Gas and Electric and Southern California Edison had appealed to the federal regulators, arguing that being forced to buy power from independent sources like wind companies would prove too costly. In addition, they said, the power wasn't needed. The Federal Energy Regulatory Commission (FERC) sided with the utilities, and Kenetech's contracts for thousands more megawatts of wind power in California suddenly dissolved, a loss that *Forbes* magazine pegged at some $945 million.

Technical issues were also surfacing with the vaunted 33M-VS, the type that went into the Delawares. At Buffalo Ridge, Minnesota, and Palm Springs, California, some blades and turbine components were failing, according to a December 1994 story in *Independent Power Report*, an industry newsletter. "Critics, who are for the most part competitors of Kenetech, have observed one-third of the turbines at the Buffalo Ridge and Palm Springs sites not operating during both light and high wind conditions," states the article, which does not carry a byline. "Some of the machines had blades missing, and blades lying on the ground were seen to have cracks in them. Further, observers have seen a great deal of retrofit activity taking place at the Palm Springs site. None would be quoted for the record."

The company downplayed the situation and sought to shift some of the blame to suppliers, even as the problems continued. In November 1995, blades from a pair of Kenetech turbines at a Spanish wind farm broke off, forcing the company to end operations at that location. A month later, Kenetech's shares were trading at less than $2, down from over $16 after the company's initial public offering in 1993, and well below the days of $29 a share in the giddy spring of 1994.

By the time the windstorm struck in January 1996, Kenetech's debt had been handed a junk bond rating by Standard & Poor's. Kenetech's treasurer told the *New York Times* that filing for bankruptcy protection remained a possibility to keep the firm going. Soon, the company faced a class-action suit by unhappy investors, who claimed it had papered over its problems. As Peter Asmus recounts, the suit asserted that suppliers of key turbine parts had warned that their warranties would no longer hold when the machines faced winds exceeding forty-four miles per hour and that Kenetech's technological breakthroughs had been overstated. Getting at the heart of the future of clean energy, the suit charged that Kenetech's management was wrong to assert that their machines could make power as cheaply as coal, at 5 cents a kilowatt-hour. Finally, the investors argued that Kenetech had not fully disclosed bird deaths caused by company turbines in California. Osborn told Asmus that Kenetech's fundamental problem was "building up $150 million in inventory of wind turbines for projects that were not yet permitted." By July 1996 Kenetech had borrowed $55 million from banks. Top company officials, including the chief financial officer, the chief operating officer, and the chief executive, stepped down, and many employees got pink slips.

"The board had decided it had created a world-beating machine," Osborn says now. It had banked on the California market, ordering 150 sets of major components for turbines. The gearboxes had a flaw, however—cracks appeared in some of the gearbox welding—and Kenetech did not have the capital to fix all of them. "When you have defects, everyone points fingers: 'This is a design flaw.' 'No, this is a manufacturing flaw.' Whatever the case, there weren't the funds to complete the repairs."

For Kenetech's believers in Austin, these were tough times. "The stock went to zero. I think I sold it before it got to 14 cents, but I didn't make much money," says King, his analogy to Enron now even more fitting. Osborn, who had left the company in late 1995, saw his stocks spiral down, too, as Kenetech shares collectively lost over $1.1 billion in value. For the fledgling Texas wind industry, and indeed for the entire country, this was a critical juncture in the development of renewables. Turbines had just grown cheap enough and reliable enough for big projects that would hopefully last for decades, and yet the largest U.S. company in the business had just gone belly-up with little warning. "We were IT in the renewable-energy business," says King. "We were big and growing bigger, talking to big utilities like a big player." When Kenetech went bankrupt, King says, it had thousands of megawatts of wind-power development still on the drawing boards.

With the demise of the company that built its only big wind farm, the Texas wind rush might have died, once again, before it started. But it was not lost on the windmen that state government had helped Kenetech's project get around regulatory hurdles. Nor was it lost on officials that they had located a new source of income: a check for nearly $30,000 landed in the Permanent School Fund in 1996. It wasn't oil money, but it was a start.

CHAPTER 6

WINDCATTERS

The fall 1998 newsletter of the Texas Renewable Energy Industries Association, the lobbying group that Michael Osborne had helped found more than a decade earlier, opened with a story that seemed like it was from another era. Under the headline "Big Spring Embraces Big Wind," the piece begins:

> *Standing atop the ridge that day was an almost surreal experience, even for those of us familiar with the nature of the land and the people. Three horsemen sat easy in the saddle next to the chuck wagon gazing out toward other ridges to the West, beyond which lay the pioneering town of Big Spring. Folks in boots, vests, hats, gingham skirts and dancehall dresses moved eagerly toward the entrance of a huge tent which appeared dangerously*

close to lifting off in the 20+ mile-per-hour wind.
The excited buzz of the crowd was accompanied by
the sounds of fiddle and guitar. And towering above
it all was the 290-foot-tall visage of a sleek, white,
modern wind turbine.

A high school band played the "Star Spangled Banner," a cowboy poet hailed the affair with verse, and the head of Texas Utilities, a behemoth that planned to buy the power, quoted Bob Dylan about how the answer was blowin' in the wind. Michael Osborne, Vaughn Nelson, Tom Foreman, and Jay Carter Sr. were all on hand to watch as the chairman of the Public Utility Commission of Texas flipped a switch to start the turbine spinning, prompting a gasp of "totally awesome" from one band member. Locals marveled at the new energy source, which looked like it was about to transform their desolate Permian Basin town, which had long made its living off oil. "The analogy overheard was 'now we can drill above the ground as well as below,'" the newsletter notes.

Indeed, the new era of drilling above the ground was well under way. Windcatters had begun to fan out across the plains and hills of Texas, looking for the best place to put what the *Christian Science Monitor* during the hopeful times of the early 1980s had called a "new kind of cash crop." One of these was a soft-spoken young man named Walt Hornaday, who had hit upon the idea of wind while in engineering graduate school at the University of Texas during the early 1990s. Hornaday grew up in a small town called Brenham, midway between Austin and Houston, known chiefly for the manufacture of Blue Bell ice cream. After spending a youthful summer doing construction work at a coal plant in Tennessee, he realized that there was "a huge amount of money" in the business of generating electricity. He wanted a piece of that, but he also wanted to do things a little differently.

The idea of wind grew on him, and Hornaday proposed writing his master's thesis about wind-turbine construction. His advisor nixed the idea. No big wind farms then existed in Texas, and Hornaday got the message loud and clear: "Quit working on these namby-pamby, flower-power ideas, [and] get serious about my career and future." So he wrote his thesis about giant cranes, and when an uncle who had retired from the air force died and left him some money, less than $100,000, he began a moonlighting operation called Texas Wind Power in 1991. The company's business model would be dismantling broken-down wind turbines, salvaging their parts (after paying the previous owner a small

6.1. *Walt Hornaday, seen here in the 1990s, tinkers with a Jacobs wind turbine that was built after the oil crisis. Courtesy Walt Hornaday/Cielo Wind Power.*

fee), and then using those parts to fix other turbines. It was the same sort of thing that an auto mechanic does with cars, and there were plenty of turbines to work on—mostly small wind-electric ones that farmers still called "wind chargers," the sort that had proliferated after the oil embargo but had fallen into disrepair.

It was hard work. Hornaday's first-ever job was trying to take apart a machine for a farmer who lived along a lonely stretch of highway east of San Antonio. "I underestimated how difficult it was to work forty feet up in the air, hanging by a safety harness, in the wind," Hornaday remembers. "And I had a crane paid by the hour waiting while I got my act together." Eventually his knees stopped shaking and he finished the job.

Texas Wind Power grew and evolved, and after Hornaday graduated, he turned his attention to it full time. He sometimes hired family, including his mother and brother and sister, and he turned to his father, a former Texaco employee, for advice. Hornaday would drive his tractor-trailer to remote ranches and hillsides and haul back as much equipment as it could carry, because the way things were going, it was necessary to "cannibalize five turbines to make one work," he says. California, where the wind boom of the 1980s had resulted in a host of machines that never worked or worked only for a few years, was a vast treasure trove.

Hornaday remembers driving out to Palm Springs, where "there would be rows and hundreds of those machines" that were largely viewed as junk, and pulling them down and eagerly grabbing their pieces, vulture-like. (On the side, he also did maintenance for Father Joe James's 1982 turbines in Lubbock, after James left and before the new guard at Saint John Neumann took them down.)

By the mid-1990s, Hornaday's ambitions had grown. The federal production tax credit signed into law by George H. W. Bush was not scheduled to expire until 1999; it cut the price of wind power by roughly a quarter and made it competitive as long as the turbines worked. Hornaday wanted to do something better than fixing small turbines, something that would make him more money and change the way electric utilities thought about power. By 1995 he had heard plenty about the Kenetech wind project in Far West Texas, and he figured he could do the same thing. He would grow another company, called Cielo, which means "sky" in Spanish, and hire men to go out into the far reaches of West Texas to persuade landowners to let him put up the giant new wind turbines. So while Texas Utilities focused its efforts on the Big Spring wind farm, Hornaday set his sights on establishing a wind farm eighty miles to the southwest, atop the mesas outside a struggling oil town called McCamey.

        In many ways, Hornaday and other Texas wind-catchers of the late 1990s had things easier than had their oil predecessors. When you're scouting for the best place to plant a wind turbine, it's fairly obvious whether the stuff you want is there or not. When Michael Osborne put up his five turbines outside Pampa, he didn't bother setting up a wind monitor beforehand. "We knew Pampa would be windy," he says. "Everybody knew it."

But placing the turbines in just any windy spot didn't work. Winds coming off buildings or tall rocks are choppy, which means the machines produce less power and endure more wear-and-tear. The gustiest winds weren't great either, as Texas energy hands had discovered when some of the turbines blew down in the Delaware Mountains. A combination of strong and steady winds was the most effective.

Among the best sites, it turned out, were the mesas of West Texas. These sites are good for much the same reason that drivers find the winds increase on overpasses: the wind speeds up as it climbs over the backs of the mesas and then picks up again as the land drops away. "The sites done in the late 1990s were on these massive mesas—400- to 500-foot drops," says Patrick Woodson, who was one of the people Walter Hornaday hired for Texas Wind Power. Everybody could tell they were good places to put wind farms.

If wind farms were to go up on the mesas of West Texas, several things had to happen. People like Woodson and his colleague Andy Bowman had to find landowners willing to lease their property to purveyors of the newfangled technology. Then, in a twist, county officials had to be persuaded to grant the project a property tax abatement, to waive or reduce property taxes on equipment in lieu of other, lesser, payments; otherwise, given the price of power, the project would not be economically viable. Finally, they had to find an investor willing to put money into the project, which meant ordering turbines from a manufacturer, paying for and erecting them.

All of these elements were tough, especially land work. In Texas, the huge, desirable tracts in the western part of the state are often subdivided among tens or even hundreds of people, some of whom may have left the state and not even know they have the title to the land. Tracking them down can be a nightmare. Hornaday recalls that once he was trying to find a man in Bowie, Maryland, who owned a sliver of West Texas land needed for a project. Letters sent to his P.O. box went unreturned. So one day Hornaday's father mailed a broom to the box, then flew to Maryland and loitered outside the Bowie post office. "He waited there for days, and finally someone emerged from the post office with a broom," Hornaday says. The man, a pharmacist, confessed that he didn't even know he owned the land and didn't want to deal with it, but Hornaday's father persisted, and they "used a notary and sold the land on the spot."

Even owners who lived on their land had their quirks. One woman, Hornaday says, "was so afraid of meeting with strangers she would walk a mile down the road and we'd meet at the gate. We'd talk about everything at the gate." Ward Marshall, who spent the 1990s as an official with a rural utility called Central and South West, describes a "Hatfield-McCoys" issue with some West Texas landowners. One of Marshall's key men called him up one day to report that he was being arrested by the local game warden for trespassing. "I said, 'Arrested for trespassing? We have a lease to be on the property,'" says Marshall. "No matter what we did, they were pissed."

Delbert Trew, a Panhandle rancher approached in the mid-1990s by

Chris Crow, a protegé of Michael Osborne's, well remembers the day the phone rang with a windcatter on the line about his 5,000-acre ridge-line parcel: "The day that they called us, my wife answered the phone and we knew nothing of wind energy. And the wind was blowing 60 miles per hour, limbs were breaking out of the trees and they wanted to know if we'd be interested in leasing our land for wind. . . . My wife told them, 'You ought to be here today. We'd all get rich.'"

Many landowners simply could not believe that anyone was coming to offer them royalties for generating electricity from a force of nature they couldn't stop if they wanted to. Often it was 2 to 4 percent of the power the project produced; that could amount to about $100,000 to $200,000 per year for a wind farm of 100 megawatts, according to Crow, who says his company, Zond, also offered a 50-cent-per-acre signing bonus (though the company sometimes had to pay a dollar).

"We'd offer them a lease, and they'd say, I won't lease it to ya, I'll sell it to ya," Hornaday says of landowners. They thought, he said, that the windmen were "crackpots." By 2012 Hornaday's company, Cielo, owned about 14,000 acres of West Texas land. Crow, whom rivals could identify from far off by the trademark feather he wore in his ball cap, also encountered the crackpot theme. "I didn't ever meet anybody who didn't want to hear the wind story, the pitch," he says. "But they didn't just grab ahold. . . . They were kind of—'we'll believe it when we see it' kind of thing."

From the landowners' perspective, though, the wind companies sometimes acted like bullies. Delbert Trew, the Panhandle rancher who had wind leases signed for fifteen years to various outfits but has never seen a single turbine go up on his land, says that the wind developers were unhappy about the notion of his large annual family reunions and his planned hunting leases. They offered to buy his land, but he refused to sell. The companies, he said, try "to dictate everything they want to do in your ranch . . . and that's just not right."

Wind developers, for their part, wanted financial help from the counties. Hornaday enlisted Bowman and Woodson, both of whom had degrees from the University of Texas at Austin, to handle tax-abatement strategies. Bowman had studied the concept at the LBJ School of Public Affairs (where he got a joint degree with the law school), and he thought he'd end up using it on behalf of companies that made conventional things like chemicals. But Hornaday overheard the two discussing the idea, and before they knew it, Woodson and Bowman found themselves far from the respectable careers their families had hoped for, shuttling back and forth to the western reaches of Texas as agents of a company that was part of an industry that didn't really exist. Tax abatements became their specialty, but they did other types of site-development work, too, often, in those early days, with little idea of what they were doing.

Bowman, who negotiated the leases, had never done anything like it before in his life. Plenty of schmoozing was needed, mingling with the locals at events such as, in one town, an annual goat barbecue where the meat was, remembers Bowman, "the scrawniest, nastiest, most horrible thing to cook or eat that you can possibly imagine." But they learned on the fly, as they had to. Woodson found himself in his first county commissioners meeting in McCamey, trying to sell them on tax abatements for the project that would eventually be named Southwest Mesa. He was right behind a man who wanted to put a spaceport in West Texas. "The judge, who's a really awesome lady out there, joked with me years later that they had a good laugh about who was crazier that day, the guy talking about the spaceport or the guy talking about the giant windmill," Woodson recalls.

As the windcatters proceeded, they faced a range of reactions—opposition, derision, or flat-out indifference—from oil companies. Oil had ruled the land for generations; it had created towns where there used to be only emptiness and made farmers wealthy even if their land was too poor to raise cattle. There was still plenty of oil during the 1990s, but the days of seemingly unlimited supply were gone; production in Texas, after all, had been falling since the early 1970s. "We ran into them a lot," says Bowman of the oilmen. "But to be honest they were sort of on the way out. . . . It was a perfect situation for a wind farm to look good, because their annual production was lower and lower and their contribution to the county tax base was lower and lower." Once Bowman and Woodson, at the bar in the Midland Hilton, overheard oilmen trash-talking wind projects and scoffing at how the wind developers "had the chutzpah to ask for a tax abatement in the county oil and gas had built," Bowman remembers. The next day, when they drove to a county hearing on tax abatements, the oilmen were there to make the case against wind and its subsidies. Bowman was ready, with rejoinders about breaks and aid for oil: "We said, 'Look, you know if we're going to talk bullshit deals, let's talk bullshit deals.'"

In many ways oil and natural gas had primed Texas for the wind boom. The wealth beneath the land had caused people to build power plants—mostly gas plants—out in the middle of the desert, and with those gas plants came transmission lines to move the power from the remote mesas to the big cities, like Dallas or Austin or Houston. Wind turbines out in the middle of nowhere could piggyback on those power lines. "So in a weird way, wind happened because of all the oil and gas activity," says Woodson.

More important, oil and gas had shown landowners that using their land to produce energy could create a hefty stream of royalties, so that when wind companies came knocking, they kept the door open a crack to listen. Wind could not pay nearly as much as oil, but it didn't make people angry and bitter the way oil companies had sometimes done. "Your average wind lease has 5 pages of . . . 'you won't speed on my property, you won't hunt on my property, you won't drink alcohol on my property, if you start a fire you'll put it out'—. . . I mean just on and on and on," says Bowman. Clearly, he adds, the landowners were tired of "dealing with the goddam oil and gas people." Later, the leases would add more provisions forcing the wind companies to clean up the land after the turbines finished spinning, by removing the towers and equipment. This was another relic of a graveyard of oil and gas equipment that had been left around the state, as well as the California wind experience of the 1980s, when too many nonworking turbines were abandoned.

It wasn't just the oil companies' behavior, it was also how the broader drilling-for-oil system worked. Some people had signed what they thought were good drilling leases, only to find that the oil company did not drill on their land but drilled on a neighbor's land instead and drained the oil from under their property, legally, without paying a cent of royalties. Others suffered the disappointment of wells coming up dry, and still others felt the pain of royalties that were plentiful at first but slowed to a trickle as the pump jacks exhausted the oil beneath their land. "The difference was that oil [companies]—they're buying their oil and depleting their resource, and wind [companies]—they're just renting wind, and you get to keep it. You're not depleting the resource," says Crow, who nearly two decades after he got started in the wind business still carries business cards that read "The Original Texas Wind Landman," because, he says, "nobody else claims it, and nobody disputes it so far." And the turbines didn't interfere with oil production, because the pump jacks could be tucked between the wind machines and still produce not a drop less oil. One wind farm, Desert Sky, dedicated by

Texas governor Rick Perry and other dignitaries in 2002, is just down the road from one of the great oil fields in Texas, called Yates, which has produced well over a billion barrels since it was first tapped in the 1920s and may have a billion or more still to go.

By themselves, the likes of Walter Hornaday and Chris Crow could never have gotten wind farms built on a scale that mattered. What Texas wind needed was money, and by the 1990s a deep-pocketed company had in fact emerged, though, oddly enough, it was based in Florida. FPL Energy, a sister to the giant utility Florida Power and Light, decided to devote itself to generating clean energy, by which it meant both natural gas and renewables. The FPL holding company had come through the 1970s energy crisis determined to diversify, and the federal tax credits for wind worked well with its financial structure to offset guaranteed long-term profits from the regular utility side of its business. In the 1980s FPL Energy's predecessor company (called ESI) had invested in and taken control of a number of California wind farms, and by the late 1990s FPL Energy had already grown into the largest wind-power producer in the country, with projects in Oregon and Iowa as well as California. Owning Texas wind farms fit right into its agenda. "When FPL brought their balance sheet to bear into that market, they were one of the first 800-pound gorillas to step into the space, which had largely before been mom-and-pop developers," says Mark Bruce, a Texas energy hand to this day who joined the company in the 2000s.

FPL Energy put up the money to build a wind farm called Southwest Mesa atop a 2,000-foot mesa near McCamey, sixty miles south of the Midland-Odessa area. Whereas the Delaware Mountains project had gone up on a remote ridge, Southwest Mesa was built smack in the middle of oil country. "The irony is obvious," says Sam Enfield, who was working in Texas at the time for FPL Energy. "Southwest Mesa is in the Permian Basin. You drive to the site, you're going by pump jacks left and right, a lot of which weren't working."

By modern standards the project was small, just 75 megawatts spanning 2,200 acres; few wind farms under construction today come in below a few hundred megawatts. Back then, though, it was the largest wind farm in Texas. An electric utility, Central and South West (CSW), had commissioned the project after getting clearance from Texas regulators to raise its electric rates by a few cents. CSW's operating territory was "just a bunch of ranchers and cows and oil fields," according to Ward Marshall, a former official there. But the utility could sense public opinion turning in the direction of renewables, and CSW contracted with FPL to buy power for twenty years from the wind farm.

In 1999, when the site began producing enough electricity for 30,000 homes, an official with the utility read a letter from Texas governor George W. Bush to those assembled at the dedication. "This worthwhile project is proof that a strong economy and efforts to ensure a clean environment can go hand in hand," Bush's letter said. To FPL Energy's president, Michael Yackira, Southwest Mesa was just the beginning. "Texas has enormous potential for wind energy development," he said, "and we look forward to developing additional sites in the state."

Building Southwest Mesa had not been easy. Construction was plagued by rattlesnakes. Hornaday recalls that after digging the hole for the turbine foundations and quitting for the night, he and his men would have to bring a shotgun the next morning to kill the snakes that had slithered into the holes overnight, before they could set the rebar and pour the concrete. The men also had to design doors to keep fire ants away from the electrical equipment. "We did a lot of Texas engineering on that first project," Hornaday remembers.

The 107 turbines used at Southwest Mesa reflected the best of Danish engineering. They were 700-kilowatt machines, more than twice the size of the ones planted atop the Delaware Mountains by Kenetech, and they were made by a Danish company called NEG Micon. Danish turbines were emerging as the workhorses of the wind business, filling the gap left by Kenetech's bankruptcy. Denmark's economy had been jolted severely by the oil-price shocks of the 1970s, and government leaders at the time, harkening back to the Danish tradition of water-pumping and grain-grinding windmills, had invested heavily in research and development, so that by 2001 some 15 percent of Denmark's electricity came from wind turbines—many of them made locally. It was from these and other European machines, growing incrementally larger, that the enormous modern wind turbines of today arose, according to Vaughn Nelson of West Texas A&M University.

The Danish turbines had three blades and faced upwind, thus definitively putting to rest the old two-bladed, downwind Carter model (though Carter Jr. still argues that the downwind machine is "more favorable because it can be flexible, like a palm tree. The flexibility is a clever way to reduce the loading, reducing the material and the weight and the cost.") Later, NEG Micon would be bought by Vestas, which remains a giant in the wind-power business. Vestas supplied the turbines for the 40-megawatt Big Spring project, which was completed around the same time as Southwest Mesa.

Southwest Mesa, because of its size, showed that wind could be practical and not just idealistic, Delaware Mountains–style. "It proved that this was a business that could make money, not some namby-pamby do-gooder stuff," Bowman says. FPL Energy, he adds, was "hugely the driving force on this. They took a lot of risk and really proved a model with that project, and that set them up for the position they have now." Under its modern name, NextEra Energy Resources, the company remains the largest renewables developer in the United States.

By the late 1990s there was another big player involved in Texas wind, one that had deeper political connections and more sway than anyone else in the business. Kenneth Lay, the chairman of Enron Corporation, had taken a liking to renewable energy. Enron, based in Houston, had originally built its business on natural gas deregulation and swiftly expanded into electricity. It had grown into a powerful company known for its savvy in energy trading, a market it to some degree created.

In 1994 Enron had dived into solar power by entering into a joint venture with the solar division of the oil giant Amoco. The idea was to build a large solar farm in Nevada, enough to supply 100,000 people with power, and indeed a few years later the combined venture received a contract to move forward. Enron reckoned that solar energy would be cheap and competitive with natural gas and coal and planned to make money in the process. The *New York Times* asked Robert Kelly, Enron's chief strategy officer, how quickly Enron would see earnings from the panels. "Now!" Kelly replied. "We're a very impatient company in terms of profits."

If Enron claimed to be bullish on the immediate profitability of solar, it also saw the new technology as a bet on the future of energy. "Ken Lay was not your typical energy entrepreneur," says Robert Bradley, a longtime speechwriter for Lay who vocally disagreed with some of Enron's decisions to invest in renewable energy. "He had a PhD in economics, he was big-picture, he had a lot of Washington experience. He was not an engineer." What this meant, Bradley explains, was that Lay's strategy involved getting people to like Enron, which would help persuade them to buy shares. And in the wake of the global climate negotiations in Rio de Janeiro in 1992, amid the drumbeat for climate action leading up to the worldwide Kyoto Protocol in 1997, being liked meant doing something green. Enron had started focusing on climate change in 1988, says Bradley, when NASA scientist James Hansen testified at a Senate hearing on the issue and, together with Al Gore (then making a presidential run), captured public attention. More importantly, global warming might create new ways to make money. Enron specialized in gas, so if coal got sidelined the company would gain. Plus, "We were very interested in carbon-dioxide trading, which was going to be the big thing under cap-and-trade," says Bradley, referring to an as-yet unrealized system of regulating global-warming emissions. "If we have wind and solar, we have credits, right. . . . We could monetize [them] by trading."

And so it was in January 1997 that Enron bought a little-known California company named Zond that manufactured turbines. Zond was a big player in a tiny field: with Kenetech's demise, it had assumed the mantle of the "largest wind company in the United States," according to *Windpower Monthly*. Zond was a rare non-European company to make it far in the wind business, and it had been supplementing its manufacturing line by developing wind-power projects, akin to what Kenetech had done. (Chris Crow was for some time Zond's main landman in Texas, and Michael Osborne had done contract work for Zond during the early 1990s, scouting West Texas wind sites—often by air, to properly test the breezes—for $25 an hour.) "Our purchase of Zond is part of a broader strategy, a vision," Enron's Kelly said at the time. "We believe that renewable energy resources will be capturing a larger and larger share of the power market within the next 20 to 25 years."

Challenges loomed. California, which had led the world in wind development during the 1980s, had been slowly drying up ever since, with wind-power output peaking in 1994. Zond had had put some turbines in California in the 1980s, though some parts broke in high winds, which was not atypical of the technology at the time. But when Kenetech went under in 1996, shaking the wind business to its core, Zond stepped in, acquiring some of the "variable-speed" capabilities that helped turbines adjust to different wind speeds and produce energy more cheaply and reliably, and propelled itself to the forefront of the wind business.

Even before Kenetech's demise, Zond was one of the early players in Texas. In 1995 it supplied turbines for a small, experimental wind farm on the grassy hills known as the Fort Davis Mountains. The project was operated by Central and South West, which would commission the Southwest Mesa project a few years later. Ed Gastineau, the utility's head of research, had persuaded top executives to put up about $10 million for solar and wind research, according to company official Ward Marshall. Gastineau had also wrangled another $7 million from the Department of Energy and a California-based utility consortium called the Electric Power Research Institute. Dabbling in wind wasn't uncontroversial—Marshall remembers one man saying, "'Why the hell would we do that? It's stupid'"—but Gastineau argued that the federal production tax credit would help wind compete on price. Part of the money Gastineau helped procure went toward buying twelve experimental wind machines made by Zond, which was eager to test different designs under different wind conditions; twelve more went to a site in Vermont.

The Texas site was not far from the McDonald Observatory, a University of Texas research center near which CSW was also operating a solar park. The turbines were made in Zond's headquarters in Tehachapi, California, and they were about twice the size of anything seen in California, says Marshall. The Fort Davis wind farm had a friendly rivalry with the far larger one built by Kenetech in the Delaware Mountains, about sixty miles to the northwest: both went online about the same time. And on January 17, 1996, the twelve 500-kilowatt Zond turbines went through exactly the same horrific storm that felled some of the larger machines on the Delawares' ridgeline.

"I'm looking at the screen, and I see wind speeds of greater than sixty miles per hour," says Marshall, then in charge of the project for CSW. Marshall had been briefing an executive team on the wind farm at the moment he saw the readings, and he figured the control system, known as SCADA, was broken; there was no way those wind speeds could be true. Plus, the sixty miles per hour figure was actually a ten-minute average, which meant that gusts were considerably higher. He called up the site supervisor, Brian Champion. "'Brian,' I said, 'I think the SCADA system is broken,'" Marshall recalls. "He goes—'Ward. It's not broken. I can't talk to you right now. I've got to take cover and get the heck off the top of this mountain.'"

The turbines survived the storm essentially unscathed, Marshall says. But over the long term, the blades fell apart, one by one. "The ailerons on the blades would just tear themselves up," Marshall says, referring to what were essentially three-segmented flaps. Many of the blades didn't even last a year. The turbines in the Vermont project, which had a different design with more rigid blades (as opposed to the flaps), held up. "Their project is still operating today, and mine is gone," Marshall says, sixteen years after the turbines began producing power in Texas.

But the idea behind the Texas project had always been experimental—to discover what worked well and what didn't in a turbine blade. Zond saw how the blades got destroyed and learned how to improve its machines, especially with the help of the new Kenetech technology. By 1997, the year Enron bought Zond, the company was producing 2.4 percent of the world's turbines. "Renewable energy will capture a significant share of the world energy market over the next 20 years," Kenneth Lay, Enron's chairman and chief executive, predicted at the time. Enron, he added, "intends to be a world leader in this very important market."

The wind world was divided about the acquisition. *Windpower Monthly* reported, "For some, the Enron-Zond deal was proof that wind power had indeed come of age and that even companies who made a great deal of money from oil and natural gas were now seeing the light when it comes to the bright future of renewable energy. For others, the purchase was a cynical attempt by a company spending $200 million in national advertising to cloak itself in the virtues of green energy, while continuing to peddle more polluting forms of electricity to the majority of its customers."

By 1998 Enron—with its new, Zond-inspired division known as Enron Wind—was ready to make another acquisition. Europe was still dominating the technology race, so Enron settled on a German company called Tacke. In a stroke Enron Wind boosted itself to become the third-largest global wind manufacturer, with a 13.5 percent market share. Tacke had built a 1.5-megawatt turbine, and with the acquisition, Enron "leapfrogged over everybody," says CSW's Marshall, who later joined the Houston office of a wind company called Pattern Energy. "They went to a turbine that was three times bigger than anyone else was manufacturing." In 1999, when the four-year-old Delaware Mountains wind farm was expanded to nearly double its size, it was forty Zond turbines that went up on the ridges, close to the old Kenetech ones.

CHAPTER 7

A WIND REQUIREMENT

On a spring afternoon in 1992, at the elite Berkeley Hall School in the Santa Monica Mountains high above West Los Angeles, about as far culturally from the West Texas plains as one can get, an eleven-year-old girl named Christiana Wyly had a moment of environmental awakening that would shake the Texas power market for a generation. She was running around the football field—Berkeley Hall today occupies sixty-six acres, with basketball and tennis courts, a seventy-five-foot outdoor swimming pool, and classrooms in a string of ranch-style buildings—and, like a young Olympian goddess, she cast an eye down at the brown haze that sat atop Los Angeles. She was suffocated by terror. "I had this feeling of powerlessness of breathing in dirty air and I could do nothing to stop it," she says.

As it turned out, she could do something about it: talk to her father, who happened to be Sam Wyly, a Republican stalwart who grew up in the Louisiana cotton fields, and with ambition, clever corporate think-

ing, and luck, had risen to become one of the wealthiest men in Dallas–
Fort Worth. Born in 1934, in the teeth of the Depression, Wyly was
raised in a clapboard cabin in Lake Providence, Louisiana, once deemed
the poorest town in the nation. His mother, unable to buy curtains or
linens, made them herself. She sold her wares to the townspeople, in
one case trading them to an Italian woman who provided "what seemed
to be a never-ending supply of homemade spaghetti and meatballs,"
Wyly writes in his memoir, *1,000 Dollars and an Idea.*

Those early trades, and calculations about how to get the best price
for the family's cotton crops, provided the primitive insights for what
would eventually become a career as an entrepreneur. Football, too,
offered some lessons. In Wyly's telling, playing undersized noseguard
for his high school football team, for a coach who was a tall Baptist deacon named Raymond Richards, taught him about the toughness and
persistence that he would later apply to business. Wyly writes about
Richards admiringly: "August in Louisiana is hot, and doing push-ups,
running laps, and doing wind sprints on a dusty field under the burning
glare of the afternoon sun were nothing short of brutal. It was so bad
one afternoon that, after running sprints for the fourth time, I collapsed
in front of Coach Richards and pleaded for mercy. I didn't have another
sprint left in me. He wasn't buying it. He said to me, 'There is no limit
to the endurance of a 16-year-old boy.'"

To most readers Richards would seem like a sadist out of *Cool Hand
Luke.* Not to Wyly. "He was teaching me that what we perceive as our
limitations are often only mental obstacles," he writes. His small size,
he eventually realized, could actually be an advantage. "Being small, I
could get down real low, and being strong, I could hit the guy real hard,
using the power in my legs to lift him up and out of the way," he writes.
"Coach Richards was the one who first told me it was called 'leverage.'"

Wyly eventually went off to Louisiana Tech, where he learned ac-
counting, and then earned an MBA at the University of Michigan. He
became a company man at IBM, selling computers for Big Blue in the
Dallas area before growing restless with the clamped-down corporate
culture. After three years he struck out on his own, ultimately starting,
in 1963, University Computing Company (UCC) with a thousand bucks
and three customers. And then he started applying leverage, building
up UCC in the face of IBM and making deals to expand his holdings to
steakhouses, oil refining, gold mining, and data transmission companies. By 2006 *Forbes* pegged his fortune at $1.1 billion.

But the question that animated him in 1992, apart from how to make vast sums of money, was the one posed to him by Christiana, who had been shaken by the thought that she could do nothing to avoid inhaling bad air. On the problem of waste, she thought to herself, we can turn to recycling. With water-quality issues, we can treat the problem. But what could be done about the toxins getting dumped into the air, she asked her father.

According to Wyly, he was already working on the problem. In 1973 the EPA began a phase-down program to reduce the levels of lead in gasoline. The new rule pushed a Wyly company, Earth Resources, whose business included oil refining, to make an expensive decision. Either Wyly could hold out for a few years, legally selling leaded fuel until the regulations took effect, or he could pay a bundle to convert sooner and hope he could sell his products for a premium. He chose the latter option, and when workers reconfiguring the refineries realized that they could also make more jet fuel and sell that at a higher profit, the early investment ended up helping the bottom line.

"I don't trust any air I can see," says Wyly, who lists the view of brown air from his beach house in California and the reading of Rachel Carson's *Silent Spring* as moments of environmental revelation. He has been known, he writes, to drive to dinner in a "caravan of Priuses" with his family.

The answer to his daughter's question was to invest in the emerging green economy. In 1997 Wyly bought a $30 million stake in Green Mountain Energy Resources, a subsidiary of a sleepy Vermont utility that hoped to break into deregulated electricity markets as a renewable-energy competitor. Green Mountain's own research suggested that a fifth of Americans were worried about pollution and would pay a small premium for an environmentally friendly electricity product. With U.S. households spending $100 billion annually on electricity, the market appeared lucrative.

Critics would eventually accuse Green Mountain of not being as green as it claimed, saying that it sold customers renewable electricity from projects that had already been built, as well as from new ones. But this did not deter Wyly, who along with his investment group became the company's principal owner. He shortened its name to Green Mountain Energy and prepared to leap into a far bigger market. "We're combining three elements that can't fail: the Vermont environmental ethic, Texas capital, and old-fashioned American entrepreneurs' frontier spirit," Wyly said. In 2000, determined to make Green Mountain a bigger player, Wyly moved its headquarters to Austin.

Just as he had competed against the monopolies of IBM and AT&T, Wyly set his sights on trying to loosen the hold of the major utilities. To do that, he would have to work to deregulate the market, because consumers would have to choose to buy his green electricity, and they couldn't while they were captive to utilities that generated and sold power without consulting anyone except state regulators about what type they were producing. Deregulation would not go Wyly's way in California, Ohio, and Pennsylvania—problems he attributed to weak-kneed politicians—but it would go his way in Texas. "Texas is more Adam Smith," is the way Wyly explained it in an interview. "Ohio and California are more Vladimir Lenin." But, curiously enough, it would be a deal that included a government mandate, right there in the heart of libertarian, capitalist Texas, that would smooth the way for a wind revolution in the Lone Star State. And it would involve a Wyly friend and frequent beneficiary of his campaign largesse: Gov. George W. Bush.

George W. Bush was never the likeliest person to jump-start the renewable-energy movement in Texas. He grew up in the oil heartland of Midland, the son of a Yankee blue blood who had moved there after serving in World War II, made a pile of money off drilling, and then turned to politics. As a boy in the 1950s, George W. Bush watched Texas suffer through its worst-ever drought, which devastated anyone making a living off farming. In his memoir, *Decision Points*, Bush would remember that the ground around Midland was "flat, dry and dusty," but beneath it lay a "sea of oil." He followed his father to Yale, where he spent plenty of time boozing and not enough studying, and after flying planes for the National Guard he ended up with a degree from Harvard Business School even though he didn't know what to do with it. It took meeting a librarian named Laura at a Midland barbecue in 1977 for him to get serious about much of anything.

He certainly got serious about oil. He'd spent a youthful summer roustabouting on a Louisiana offshore rig, and in 1979, after a stint as a Midland landman, Bush set up an oil and gas exploration business of his own. He was drilling oil wells, some of them dry, when crude prices began their sudden collapse in the 1980s. His father, George H. W. Bush, was vice president under Ronald Reagan by then, and W. had a front-row view of the administration that Michael Osborne would blame for killing off the renewable-energy industry.

By the time he beat out Ann Richards in what the *New York Times* deemed a "stunning upset" to become governor of Texas in 1995, Bush was a big believer in the free market. Government tended to get in the way, in the environmental field and everything else. He styled himself as conservation-minded, but during his governorship, he would repeatedly infuriate green groups, especially on air pollution, which was worse in Texas than just about anywhere else in the nation due to the near-ubiquity of heavy industry and a reluctance to force factories to clean up. When Bush was running for president in 2000, one leading environmentalist told PBS *Newshour* that the governor had shown "a great deal of indifference to the environment," and another that he had "stood up for the polluters rather than the people" at every opportunity.

Given this record, what prompted an odd exchange with his top electricity regulator in early 1996 is unclear. Perhaps just one year into his governorship, Bush had begun to dream about the national stage, the one occupied by his father for only a single term. Or perhaps he was thinking back to his years in Midland, the place where he met and married Laura, the place where West Texas winds blew hot and dry and carried swirls of dirt over the oil fields. Bush, as an oilman, would have known that the small towns around Midland faced shrinking production and would welcome the jobs a new industry could bring.

One day in 1996, Bush's thirty-three-year-old chairman of the Public Utility Commission, a good-natured energy wonk named Pat Wood III, was heading out the door of the governor's office when he was stopped short by some unexpected words. "Oh, Pat, by the way, we like wind," Bush said. "We what?" Wood stuttered, dumbfounded. "Go get smart on wind," Bush replied.

Whatever the motivation, the approach was classic Bush. The governor and future president was nothing if not a big-picture guy, articulating a vision and letting others work out the details, mostly. "I'm always the first to admit that I always rely upon smart people and they're smarter than me," he told a reporter in October 1996 during an interview about Wood and other PUC commissioners.

In Wood, who much later would follow him to Washington to become chairman of the Federal Energy Regulatory Commission, Bush had found a faithful lieutenant, whose combination of likability and savvy would help him push his priorities through the byzantine world of Texas politics. Wood was a former Eagle Scout who spoke Spanish and German, and with his mop of hair and round face he still looked boyish. He had studied civil engineering on a Texas A&M University scholarship and then attended Harvard Law School, where he was elected student body president and graduated in 1989. But he had won an earlier, more deeply felt education at the drugstore that belonged to his father in Port Arthur, in the heart of Texas's oil-refining coastline. Wood's great-grandmother had opened the store in 1926, and as a boy he had learned a thing or two about customer service and competitive pricing. In 1993 an Eckerd drugstore, part of a nationwide chain, moved in across the street. Wood witnessed his father wage a valiant and, in the end, successful battle to stay in the drugstore business as he slashed prices and remained as loyal to his customers as they did to him. In Wood's mind, competition fostered positive results.

Growing up in Port Arthur, where oil was a way of life, also got him interested in energy. His grandfather had managed Gulf Oil's flagship Port Arthur refinery. And just hanging around Port Arthur, watching men go to work in the refineries and hearing about how his great-uncles had piloted tanker ships, Wood had a glimpse of his future. "That was the town where oil was discovered," says Wood, "so that was what I always wanted to do."

Wood's path to the energy regulatory world would be impressively direct. In between Texas A&M and Harvard, he had worked for an oil company called Atlantic Richfield (better known as ARCO) until the oil market crashed in 1986, and after Harvard he spent a brief stint at a law firm before heading to the Federal Energy Regulatory Commission in Washington. At FERC he served as general counsel to a Democrat, Jerry Langdon, appointed by Pres. Ronald Reagan. If there were exciting times to be an energy regulator, these surely were it: George H. W. Bush, who succeeded Reagan, was pushing forward with the restructuring of the natural gas industry, which needed it. Producers of natural gas in Texas and elsewhere sat on plenty of supply, but for complex regulatory reasons few companies in the Northeast wanted to buy it because prices remained stubbornly high. To remedy this, regulators first took away the monopoly power of the natural gas pipeline companies, so that any gas producers could send gas through them far more easily (this was the way oil pipelines had worked essentially since Spindletop, but gas pipeline companies, curiously enough, had not). Thus, gas producers began to be allowed to sell their products directly to their customers, rather than having to rely on the all-powerful pipelines, which previously had not only moved gas but also bought and sold it, as a marketing middleman. "I loved it," says Wood. The idea of restructuring and opening a closed and stubborn industry to increased competition made perfect sense to the self-described "young Reaganaut," who was in the thick of it in his job in Washington. "It was just intellectually something I bought into," he says.

When Wood moved back to Texas two years later and took a job as legal counsel for one of Texas's three railroad commissioners, who headed the state agency that regulated oil and gas and a few other industries (though no longer, really, railroads), the notion of deregulation and competition was expanding into new industries. When a new, Republican governor arrived in office with an oil and gas background and an interest in deregulation and free markets, Wood's network kicked into gear. Three of Wood's former bosses told Bush he had to talk to Wood, and Wood went into the interview with humorous resignation. "I'm like, 'Oh, man, I've been working for the government for four years, my dad can't hold his head up straight because his son's working for the damn government,'" Wood recalls thinking. Whatever his internal angst, Wood would have a few more years with the government. Bush, impressed, appointed the young man on the spot to serve as one of his top energy regulators.

When Bush casually called out to Wood to "get smart on wind," Wood was first incredulous, then dubious. As far as he knew, wind energy was not competitive with the cost of fossil-fuel generation. And even if it cost only fractionally more, would consumers really care enough about pollution to pay the premium? When Dale Osborn, the head of California-based Kenetech, was making the rounds in Austin a year or so earlier, Wood, as chair of the Public Utility Commission, had cut off any suggestion that the government would help foot the bill for such projects. "I was not Mr. Tree Hugger," Wood once told the *Austin American-Statesman*. Indeed, when his boss called out to him that day, Wood recalls this thought flashing through his head: wind, as far as he knew, was "California, Volvo-driving, Birkenstock-wearing, tree-hugging kind of stuff."

But Bush was serious, and others noticed, too. A spring 1996 newsletter from the Texas Renewable Energy Industries Association carried the headline: "Bush to Agency Heads—Pay Attention to Renewable Energy." The story was about a new document that, among other things, urged government officials to stay attuned to environmental matters, including paying attention to the percentage of renewables on the grid. It wasn't much, but it was a start. "The governor is to be commended" for his vision, the renewables group concluded.

Soon Wood was serious about wind, too, thanks to a sort of renewable-energy epiphany that came through an unexpected channel: public opinion polls. In 1995, with an eye to the kind of restructuring of the electricity industry that captivated people like Wood and Sam Wyly and Enron's Kenneth Lay, the State Legislature ordered electric utilities to undertake a process cumbersomely called "integrated resource planning." Lawmakers wanted utilities that needed more power to start shopping around for the best deal, rather than just resorting to the usual strategy of building more coal, natural gas, or even nuclear plants. This could take the form of buying power from non-utility sources or encouraging customers to use less electricity. As part of a 1995 law, the legislature required planning that included a look at renewable energy and energy-saving programs. It also required public participation in creating the plan.

And so it was that between 1996 and 1998, eight Texas electric utilities embarked on the humdrum-seeming task of polling their customers to figure out how they wanted to meet the future electric needs of a growing state. This wasn't a run-of-the-mill telephone poll. Conventional polls on arcane energy matters would be useless, since the companies "knew that the public did not have the information, or even opinions about the issue worth consulting," James Fishkin writes in *When the People Speak*. A focus group might not be representative of the public, he explained, and a town meeting risked getting overrun by organized advocacy groups. Furthermore, says Dennis Thomas, who served at the time as a consultant for Central Power and Light, a subsidiary of Central and South West, the companies wanted to make sure the polling results they assembled "were bullet-proof in terms of methodology" to defend against inevitable legal challenges. Thomas approached Fishkin, then a University of Texas government professor, to do what Fishkin had trademarked as "deliberative polling." The idea behind the polling system was to determine not just public opinion, but *informed* public opinion. The core question animating a deliberative poll, as a paper on how it worked in Texas explains, is, "What would the target population think if they were given an opportunity to read about, discuss, and ask questions concerning the issue under consideration?"

"It was a marriage between scientific survey and New England town meeting," says Ron Lehr, a Colorado lawyer who helped organize the polling sessions. They worked this way. The utilities asked a representative sample of citizens to join an exercise for one or two days, generally on a weekend. Every effort was made to make participation easier. "If they say 'Yes, but I need a babysitter,' then you buy them a babysitter," says Karl Rábago, a former public utility commissioner in attendance. The organizers then gave the participants information (vetted by a diverse advisory group) about energy issues in Texas. At the event, the citizens would discuss the issues in both large and small groups and could pose questions to regulators and utility executives. "Rich, poor, men, women, different backgrounds," Lehr says, "people start to work on each other."

At the end of the exercise a poll would be taken to see if, and how, opinions had changed compared with a similar poll at the beginning. In Texas the verdict was clear and astonishing. "Taken together, renewables and efficiency are clearly preferred by most customers after the event, while coal, natural gas, and purchase power are less preferred," stated a 2002 retrospective report on the project for the National Renewable Energy Laboratory. And Texans were willing to spend more on renewable energy. "In our observations of almost 100 small group discussions, we heard participants repeatedly talk about renewables not just as a non-polluting resource, but one that could eventually become our primary source of energy," wrote the report's authors. "This suggests that people look at renewable energy as having both a present and future value." Specifically, the pollsters found, the percentage of participants willing to pay more on monthly utility bills to support renewable energy—from $2 to $5 more, according to Thomas—jumped from 52 percent to 84 percent, on average, over the eight sessions. Fishkin, in an interview, calls a shift that large "an earthquake in public opinion."

One panelist at the first deliberative poll, in 1996 in Corpus Christi, was Pat Wood. The previous year, in a March Public Utility Commission newsletter, he had expressed his regulation philosophy this way: "With a close watch on the best interests of all Texans, I would like to see the market replace the regulator where possible." The Public Utility Commission, he felt, should serve "as a policeman of market forces rather than as a substitute for them," as he wrote in a June 1995 memo to another commissioner.

It was with this mind-set that Wood arrived at the Corpus Christi deliberative poll, along with his two fellow public utility commissioners. Karl Rábago, who attended as a representative of the Environmental Defense Fund and thus felt free to push an environmental agenda, got to work. Rábago recalls sauntering out during lunch hour in Corpus Christi to fly a kite in a nearby field—subtly hinting at the strength of the wind. "[People] were going, 'Shit, it is kind of windy here,'" he says. Wood remembers giving opening remarks to the roughly 250 citizens assembled, then watching through a window as small groups of about fifteen chewed over problems like whether it was desirable to buy power from the open market, rather than having it all bundled up and sold as a package by utilities; whether energy conservation or building new power plants was better; and whether renewable energy was a good idea. Any questions they had, they could put to the experts. Fifteen years later, Lehr still recalled what happened then with a sense of wonderment:

> *I tell them to shoot the hardest questions they can.*
> *Wood's on stage with his lawyer, who's protecting*
> *him or something. He was new. Some guy stands*
> *up, and says, "I want to make an observation that*
> *I think there should be a statute in Texas encour-*
> *aging renewable energy." Pat says, "I'm a new Bush*
> *appointee. We're Republicans, we're market ori-*
> *ented. If there's demand there will be supply. You*
> *will be well served by the market." The guy walks*
> *away, kinda shaking his head, and I say, "It doesn't*
> *look like you're satisfied with the answer." He turns*
> *back and slams his hand down on the podium. "We*
> *pay you big bucks to go to Austin and you're not*
> *going to do anything." And there were probably 250*
> *people in the room and you could hear them all go,*
> *"Grrrr." Pat Wood blinked and said, "If that's how*
> *you feel we'll have to do something about it."*

The truth was, Wood came away from that meeting amazed at the eagerness for renewable energy in Texas. "The switch in support for conservation and renewable energy was just dramatic," he recalls. "It really opened my eyes." More remarkable still, the clamor was not coming from "people in hoity-toity Austin, or, God forbid, people from the East Coast or West Coast." They were people from Corpus Christi, a port town serving the petrochemical industry, and they wanted big changes—changes that would green up the electric system. "I was like, 'Oh, that must be a mistake.'"

It was no mistake. The experience, repeated once in Louisiana and six more times around Texas, in places like Dallas and Amarillo and

Beaumont where there were no hoity-toity Austinites either, showed almost the same result. People wanted renewable energy, and they wanted their policy makers to start procuring it. So when George Bush asked Wood to work on wind, he ultimately was glad to do so, and a few years later he traveled to the West Texas hamlet of Big Spring to turn on the mighty wind machine that would create power for Texas Utilities. "Texans have overwhelmingly told their electric companies that they want more renewable energy because they want clean air," Wood told the crowd that day in 1998. But wind was not the only thing that Texas ratepayers wanted—or were going to get.

In a grand irony, Texas lawmakers tied a mandate that utilities invest in renewable energy to a massive effort to deregulate the electricity market. In other words, to use Sam Wyly's language, Vladimir Lenin somehow sneaked his way into an Adam Smith cocktail party. The irony seemed largely lost on the players at the time, partly because the renewable mandate was a small part of a massive process of horse trading. To get utilities to invest in renewable energy, among other things, you had to give them a lot in return.

The urge to restructure the electric market was part of a lingering Thatcher-Reagan-era shift that supposed, broadly, that customers were better off if any given industry faced ramped-up competition. Deregulation actually began before Thatcher and Reagan took office, as the Chicago school of economics gained devotees in Washington. Presidents from both parties took up the call for freer markets. The Railroad Revitalization and Regulatory Reform Act of 1976, signed by Pres. Gerald Ford, aimed at improving train service, especially in the Northeast. In 1978 the federal government deregulated the airline industry, accelerating competition, dismantling controls of airfares and routes, and giving upstarts like Southwest Airlines in Dallas a runway to success. That was also the year that Congress passed the Public Utility Regulatory Policies Act, affectionately known as PURPA, which pried open the monopoly power of electric utilities just a crack by forcing them to pay small providers like Michael Osborne in Pampa or Father Joe James in Lubbock a fair price for the electricity they produced for the grid. The year 1980 saw the effort to deregulate the trucking industry, and after a federal judge signed off on the breakup of AT&T in 1984, Americans for the first time could choose what company would supply their long-distance service. Wood would work on some of these issues later, under Governor Bush, in addition to the ongoing efforts to remove constraints on the natural gas market.

Texas would not become the first state to deregulate its electric sector. By the late 1990s eighteen states were already in the process of restructuring their power industries, so that Goliath utilities no longer monopolized the business of generation, transmission, and selling of electricity. Texas began this work in 1995, when the legislature, in a move dubbed partial deregulation, allowed utilities that generated power to choose what company would buy their power and sell it to consumers rather than just being stuck with one option.

But deregulating the Texas market was hard. For one thing, history was against Wood and Bush. The roots of Texas regulation went as far back as 1921, when the State Legislature decided to create a single system for generating and transmitting power. Before that, according to a January 1997 edition of the *Perryman Report*, a newsletter by Texas economist Ray Perryman, utilities built their own power plants and wires, resulting in an uncoordinated, ad hoc system in which work was often duplicated and some areas remained unserved. Electricity flickered on and off more than anyone liked, and rates remained uncomfortably high for the average homeowner. (According to that 1997 *Perryman Report*, Dallas customers in 1917 would have paid $1.17 per kilowatt-hour in 1995 dollars, whereas in 1997, they would have paid far less—about 6.5 cents—for a better product.) Rural customers, of course, had no hope of centralized service before Franklin D. Roosevelt's rural-electrification push.

Regulation through the years stabilized prices and ensured that the lights stayed on, and the electric utilities grew in size and clout, serving customers within a specified geographic area. By 1995, when Wood was appointed, the Public Utility Commission was overseeing utilities that earned about $15.5 billion a year and provided power to 6.8 million customers. So why would lawmakers want to change that system? "The public utility system we have now works to benefit all Texans by promoting economic growth and prosperity. Further change in the industry would likely impose uncertainties on a system that is producing valuable economic benefits for all Texans," Perryman concluded.

But the utilities and their biggest customers wanted deregulation badly. Those customers, which included large manufacturers and refiners, thought competition would lower their monthly bills. And the utilities wanted access to new markets, beyond their conventional, geographically delineated service territories like Houston or Dallas. Even some environmental groups and consumer advocates could get behind a measure if they could be convinced that deregulation would be a means of cutting power plant emissions or getting more renewable energy into the grid.

Texans didn't exactly have sky-high rates—compared to Hawaii, the most expensive state, where customers paid 14.7 cents per kilowatt-hour for power, Texans in 1998 paid far less, just 7.8 cents, which was about the middle of the pack for the nation. But with high summertime use of their air-conditioners, Texans were stuck with the third-highest average annual electric bills in the country. To be successful, the deregulators would have to make assurances that bills would not go up, at least not in the short term.

Bush wanted a deal, too. As early as 1997, after he had invested in Green Mountain, Wyly checked in with the governor. Bush's "number one interest was a competitive marketplace for the businesses that are the big buyers of electricity," says Wyly, who had become a chief backer of the governor. (Along with his brother, Charles, he gave Bush's 1994 and 1998 gubernatorial campaigns $210,000 total.) A competitive marketplace is exactly what Green Mountain wanted, too, so that it could sell its renewable-energy products across the state. A 1997 effort to further deregulate the electricity market failed in the legislature—it started late and didn't even get out of committee—after electric cooperatives, among others, balked at the prospect of competition and, they feared, higher costs for power. Just about every lawmaker had a cooperative in his or her district, so they acceded.

But by 1999 Wood had found some key partners in the legislature willing to take on the grand bargain. One was Steve Wolens, a Democratic state representative and former trial lawyer from Dallas whom *Texas Monthly* would describe as an "intellectual gladiator." The other was David Sibley, an imposing Republican state senator from Waco who had once played basketball at Baylor and worked as an oral surgeon before an accident when he was just thirty-six forced him to change careers and attend law school. They were among the brightest lawmakers at the state capitol, and each chaired an influential committee—Wolens, State Affairs; Sibley, Economic Development. Both remained determined to work through deregulation after the 1997 effort failed.

The way Steve Wolens tells it, a journey he took to the Magic Kingdom helped turn wind power from a fringe fantasy of Texas hippies and off-the-gridders into a mainstream, corporate reality.

In 1994 Wolens was on vacation, hauling his wife and three kids from La Jolla to Disneyland, when he spied giant wind turbines looming along the hills above the San Diego freeway. He had never seen anything like that around Texas and he bookmarked the image in his mind. What Wolens saw during that trip to Disneyland drove him to include legislative language that would finally spur the genesis of a wind industry in Texas. "It was something that I wanted," he says, and it didn't hurt that it would corral the support of environmentalists, who had a surprising amount of political clout.

Sibley, for his part, felt that the promise of renewables was "worth trying," but he was worried about the cost. His cooperation proved crucial. "Sibley had clear command of the senate," the energy reporter Bruce Hight wrote in the *Austin American-Statesman* on March 18, 1999. "Without exception, if he favored an amendment, it passed; if he didn't, it failed." Sibley was also concerned about yet another moving part in the ambitious deregulation effort—cleaning up the state's oldest and dirtiest power plants. Some coal plants were near Wolens's Dallas district, but Sibley also made clear that he did not want jobs to be lost at an old but dirty natural gas plant called Tradinghouse near Waco, according to Tom "Smitty" Smith, the head of environmental group and government watchdog Public Citizen. Sibley was placated on both fronts, Smith says, only after seeing a study commissioned by Wood's regulatory shop showing that pollution controls and wind-energy projects would add up to less than $2 per month per electric customer. That was when, according to Smith, Sibley said, "I'll put it in the bill."

The 1990 Clean Air Act, which established a federal system for capping emissions of nitrogen oxides and sulfur dioxides and allowing companies to trade pollution-reduction credits, was supposed to have forced a clean-up of many of the most polluting coal plants. But environmentalists complained that while it did ease the problem of acid rain in the Northeast, it had failed to shutter the plants, which kept belching pollution. Texas, as a large and heavily industrialized state, had more than its share of pollution. So Texas green groups like the Environmental Defense Fund and Public Citizen and the Sierra Club wanted to force companies operating in the state to use cleaner technology by capping the pollution they were allowed to emit.

Bush favored voluntary clean-up of the old coal plants. That was his philosophy: government could guide but should ultimately get out of the way. But Wyly, whose daughter Christiana had drawn his attention to dirty air, pushed for mandatory compliance. Smith remembers talk-

ing to Wyly perhaps ten times during that crucial 1999 legislative session; he would relay any problems environmentalists were having, and
Wyly, who "clearly communicates at a level that none of us could deliver
with Bush," would talk to the governor. In an April 1999 e-mail to Bush,
Wyly wrote, "Texans need guaranteed and significant improvement in
the air we breath[e]. As I have stated on many occasions, the stars at
night, are no longer big and bright, deep in the heart of Texas. I feel
strongly voluntary guid[e]lines to clean up our air simply will not work.
I urge you to stand firm and accept nothing less than a mandatory 50%
reduction in emissions by 2003. Also, I support the proposed cap and
trade system."

Environmentalists were hard at work, too. Chuckling with pride,
Smith remembers an event in Dallas in which a colleague from Public Citizen stashed a Tickle Me Elmo doll in her purse, and held it up
against her armpit so that the doll would "laugh and giggle" whenever
Bush, who was speaking at the event, said the words "voluntary reduction": "The first time that happened Bush was taken by surprise. He
kind of looked around. The second time he was going to say 'voluntary
emissions reduction,' he was a little bit cautious, so she did it again. The
third time, he could barely get the words out. . . . By the end of the press
conference everybody was looking round and laughing every time the
words voluntary emission reductions came out of Bush's mouth." With
every snicker that echoed through the hall, Smith grew more hopeful
that environmentalists could turn the tide in favor of cleaning up coal
plants—and perhaps point momentum toward a new technology that
emitted no noxious chemicals into the bargain.

In the halls of the Texas capitol, there's an old say-
ing: The legislature is designed not to pass bills, but to kill them. Texans don't like too much government, and it shows in how the legislative
process works. The 1876 constitution, the one currently in force, contains a key phrase, untouched through the years: "The Legislature shall
meet every two years, at such time as may be provided by law, and at
other times when convened by the governor." An amendment passed in
1960 specifies that even when lawmakers do convene in Austin, they
shouldn't let their work take more than 140 days. So whereas California
lawmakers meet year-round and can pass legislation at any time, Texans must cram it all into less than five months, every two years—"such

a brief window that it's almost breathtaking," in the words of California lobbyist V. John White. If the deadline passes, the legislation dies.

To pass a massive, complicated bill like electric deregulation therefore required plenty of advance preparation. Getting the law right meant lining up a kind of Rubik's cube of interests, from investor-owned utilities to organized labor to environmental groups to large businesses that were the chief electricity consumers. And so much of the heavy lifting on deregulation began well before the legislative session even started. As soon as Bush took office in 1995, he began hearing from old friends like Kenneth Lay of Enron, who were clamoring for deregulation. Lay wrote frequently to Bush on stationery bearing his company's famous crooked E. His secretary's "Dear Governor Bush" would be crossed out, and Lay would pencil in "Dear George" instead. He wrote to invite Bush to musicals, to commiserate over knee surgery, or to express gratitude for a Christmas present, but just as often he wrote to the governor about the "benefits of competition" that deregulation would bring. "We have already glimpsed this energy future, and it works," Lay wrote to Bush in 1996, a year after Texas lawmakers had begun to dismantle the electric utility monopolies. In 1994, shortly before Bush took office, Lay had even written to "Dear George" to recommend Pat Wood for the Public Utility Commission job.

As momentum for deregulation built, the trio—Wood, Sibley, and Wolens—began to hash out the electric future of the state. On a flip chart with permanent marker they diagrammed the future of Texas's electric markets. Back and forth the trio went on the diagram, Wolens remembers. They took trips together to places that had recently deregulated their electric markets, like Britain and California and Pennsylvania. Wood remembers flying in 1998 on a small plane from Harrisburg, the Pennsylvania capital, to Dulles Airport near Washington, D.C., and watching Sibley and Wolens map out a key part of Texas's deregulation package—the "price to beat," a rate the old electric providers would have to maintain—on a napkin. As the 1999 session began and the meetings intensified, Wood remembers it being "like crashing for the final exam with your study group for three months." His wife, Kathleen, was pregnant with their first child. One day Kathleen received a call from the governor. "'Honey, thank you for loaning him to us. He looks like hell,'" Bush said, as Wood recounts. Wood had a more positive memory of this time. "Intellectually," he says, "it was just heaven."

Senate Bill 7, the prosaic name for a sweeping law that launched retail energy deregulation, birthed wind power, and led to a clean-up of coal plants, was introduced on January 20, 1999, barely a week after the session started. Sibley was the official author, and Wolens—who had introduced a similar bill in the Texas House more than a month earlier—was the sponsor. "When they came out with a bill, it was a fairly well-backed bill," says Julie Blunden, who was working in Austin for Green Mountain, Wyly's project, at the time. "It was not a kind of 'Throw it at the wall and see what everybody has to say' kind of thing."

With much of the difficult negotiating work largely done before the session even began, the lawmakers and Wood were able to navigate through the shoals of a legislative process meant to shatter the hulls of the best-intentioned bills. Most of the focus was on electric deregulation, the massively complex stuff that Sibley and Wolens and Wood had mapped out on napkins and flip charts, but it also contained the following phrase: "It is the intent of the legislature that by January 1, 2007, renewable energy technologies shall constitute not less than five percent of the installed electric generation capacity that is physically located in the state and available to sell power at wholesale or retail." The Wolens bill, introduced in December, contained a goal of three percent renewable energy by January 2005.

Environmentalists went to work. They were acutely aware that these were goals, not mandates, and how those goals could possibly be reached voluntarily was unclear. Sibley's bill contained an interim requirement that by 2004 electric companies get at least 1 percent of their energy from renewables. To describe how renewables would suddenly leap from 1 percent to 5 percent by 2007, the bill cited "reliance on market forces alone." In other words, the theory was that once consumers could choose among different electricity options thanks to deregulation, they would gravitate toward renewable energy. The polls, after all, had shown that Texas consumers valued cleaner electricity, and the production tax credit generously provided by the federal government would help to keep the prices down. Environmentalists, however, worried that when it came to actually reaching into their wallets, Texans would probably go with the cheaper, more polluting conventional sources.

For his part, Bush—for all his comments to Wood three years earlier—was reluctant to commit to a mandate. "Governor has *not* staked out a position on the . . . requirement that some energy be renewable," reads a handwritten note dated February 1999, scrawled on a draft "robo," or form, letter to constituents who urged more action on clean air. In April the same year, a letter from a wind company representative to John Howard, Bush's environmental advisor, thanked Howard for agreeing to read remarks by the governor at a ceremony at the capitol showcasing the progress of West Texas wind-power development, with wind equipment in a trailer. "[D]id *no* such thing—will *not* speak" is scrawled on the letter. Undeterred, FPL Energy, which was just completing its wind farm at Southwest Mesa, near McCamey, trucked in a seventy-eight-foot wind blade, parked it in front of the state capitol one April day, and called a press conference. The blade was "almost the equivalent of a basketball court," the announcement said, and it would show lawmakers, the executives thought, just how big wind, quite literally, was getting.

Kenneth Lay, too, was against a mandate. In wide-ranging testimony on deregulation before Wolens's committee in April 1998, he told lawmakers that his company was "moving very aggressively into renewable energy" with its wind acquisitions and solar plans. Between July 1998 and June 1999, in fact, Enron Wind was responsible for installing more than half of the new wind capacity in the entire country. And yet, Lay said, "I would strongly urge that you not consider any mandates." Past rules about which fuels could and could not be used to generate electricity—like the 1970s-era bans on natural gas power plants—had created problems. "The market can still make those choices better," Lay said. Then he softened a bit. "But there may be some [*sic*], again, a public interest reason to provide some incentives for renewable energy." He did not specify what those would be.

Wind companies were dead set on a renewable-energy mandate and decided to pool money to help their cause. It would not be easy. Mandating anything was a tough sell in Texas. In 1993 wind boosters had tried to sneak a 2 percent renewables requirement into a different bill, but that had gone nowhere. They could plead their case with West Texas lawmakers, who had seen how wind farms could spread across their districts and be a good source of jobs. V. John White, a leading California renewables advocate, arrived from Sacramento at the invitation of Smith and others to help organize a lobby group. He mostly stayed behind the scenes, because "it's like the salsa commercial," says White. "You're from California. We don't want that!"

Five energy companies—including Enron and FPL Energy—kicked in about $10,000 each, in Smith's recollection, to form a lobbying coalition of renewable-energy and environmental groups. They hired three men to build their case. One was Bob King, who had ridden the Kenetech roller-coaster and helped build the first big wind farm in Texas; another was Todd Olsen, a former employee of Karl Rove's who had

not lobbied before and hasn't since, and years later described the experience as "very, very fun," a favorite in decades of working around
the Texas capitol. The third was a man named Ed Small, who stood
six-foot-four and had played tight end in the 1960s for the University
of Texas football team ("You knew he was the real dude," Smith says).
Small was a longtime lobbyist for an influential cattle-raisers group and
had often crossed swords with the likes of Smith when he represented
hog farms that wanted environmental permits. So he was surprised
when he walked into a room and saw his new allies. "I wasn't known as
somebody who worked with those environmental groups," Small says.
He went to Sibley and told him he was working with the environmentalists to push renewable energy. "He said, 'You're OK,'" Small remembers.
A signal like that meant exactly what it sounded like. The legislature's
master craftsmen needed to be left alone, and wind would get its place.

        In the final reckoning, after all the wheeling and
dealing and debate, wind was a seeming afterthought. This was one
of Texas's most sweeping bills in recent memory. It boldly deregulated
electricity markets and required some clean-up of the old coal plants
that had been bothering environmentalists. The legislation broke up
existing electric utility companies into power-generating companies,
transmission and distribution utilities, and companies that sold power
to consumers, and would allow the last group to compete to serve
homeowners and businesses. Environmental groups were won over to
the notion of deregulation with assurances that rates would stay flat or
decrease for several years. The rural cooperatives that had helped crack
the alliance behind deregulation only two years earlier were satisfied
by language in the new proposal that would exempt their service areas,
along with those of municipally owned utilities like those in Austin or
San Antonio, from competition. The Public Utility Commission, Wood's
agency, would be in charge of writing more specific rules for how the
grand plans written in the bill would actually work.

So inconsequential was the renewable-energy requirement's place
within this masterwork that Sibley's floor statement on the measure on
March 17, 1999, the day the bill finally passed the senate and got sent
to the house, didn't even touch upon renewables apart from a general
mention of cleaning the air. Wolens's office seemed similarly indiffer-
ent, mentioning the renewables requirement in only one sentence on

7.1. *Gov. George W. Bush signs the 1999 electricity deregulation law that set the path for major development of wind power in Texas. He is flanked by state representative Steve Wolens of Dallas, to his right, and state senator David Sibley of Waco, the architects of the measure. Photo by the Associated Press.*

the third page of an April 6 press release about the bill. Another Wolens release, a month later, again didn't mention renewables until the third page, after touching on wonky issues like customer choice, start date for competition, electricity prices, a five-year rate cap, an energy rebate, and rules on cleaning up old generating plants. In the end, says Sibley, there was little pushback from other lawmakers on the renewables provision. "There were a lot of moving parts and a lot of whiny people," says Sibley. But when it came time for the wind piece, a minuscule price for the utilities to pay for freer access to potential markets, there was not much whining.

Environmentalists and wind companies were quietly content. After a hard fight, they had managed to secure a requirement—an actual mandate—for 2,000 megawatts of additional renewable energy in Texas by 2009. (Besides the renewables requirement, the bill also instituted a requirement for energy efficiency, one of the first in the nation.) "Renewables" was understood to be wind because it was cheaper than solar or geothermal power, especially when the federal tax credit was factored in, and there was little chance that Texas would build more dams. Texas at the time had only 800 megawatts of renewable energy, much of it hydropower from New Deal–era dams, supplying 1.3 percent of the electric grid's peak demand. Wind accounted for less than 150 megawatts, a figure that included the expanded Delaware Mountains wind farm and the Southwest Mesa project near McCamey, which got turned on in May 1999.

The 2,000-megawatt number, far from being random, was designed to feed perfectly into good old Texas boosterism. At the end of 1998 just 1,770 megawatts of wind were installed across the United States, and as of 1999 wind accounted for only 0.1 percent of U.S. electricity production. "We wanted to be able to say that this would double the amount of wind generation in the country," says Jim Marston, who headed the Texas office of the Environmental Defense Fund and was closely involved in the negotiations along with his deputy, Mark MacLeod, who had formerly worked for the Public Utility Commission and for a study team on electric deregulation led by Sibley. It would amount to about 3 percent of Texas's electricity requirements in 2009, according to *Windpower Monthly*. (The 2,000 megawatts of capacity translates to less than it sounds like; capacity simply means how much the units would produce if they were working at their maximum all the time, which is never the case because the wind comes and goes.) Renewable-energy companies had wanted even more, but Marston figured that later on, as technology costs came down, the number could always be raised. And there were teeth in the bill, which said electric companies would face penalties if they blew off renewables.

Bush ended up backing the policy. "The governor was all for it," says Sibley. "I remember getting him on the phone from my office and I told him about the [renewable portfolio standard], and he said that sounds like a great idea." He had been hearing about wind from his friends— not just from Wyly, who wanted cleaner power generation and a market for Green Mountain, but also from Kenneth Lay. In 1998 Lay had sent a "Dear George" letter noting the "great wind potential" of Texas. Enron Wind, having bought Zond the previous year, had become, Lay wrote, "one of the largest wind turbine manufacturers in the world, [and is] spending approximately $80 million for wind turbine components in Texas." Lay may not have supported a mandate, perhaps out of a general fear that mandates could harm his other businesses like natural gas, but he told Bush that wind "will offer long-term economic and environmental benefits not only to Texas and our country, but to the world as a whole."

Between Lay and Wyly, "Suddenly you had two of Bush's biggest donors . . . saying to Bush, 'you ought to do wind,'" says Smith. Wyly's own recollection more or less concurs: "It wasn't so much that George the Younger wanted to make the state greener, although that was okay with him, in part because the green Wylys were long-term, clannishly loyal backers of his dad and him," Wyly writes in his memoir. "More important to him was using deregulation to create a competitive market in electricity."

And so it was that on June 18, 1999, six days after declaring he was a candidate for president of the United States, Gov. George W. Bush signed Senate Bill 7 into law and became the unlikely hero of the wind industry, in Texas and the nation. Deregulation was the main agenda item, but the law also propelled Texas's wind-power production past California's and that of every other state in the country. Sibley's press release that day buried its mention of renewables. But factories in Europe suddenly found themselves cranking out wind turbines and shipping as many to West Texas as possible. The growth would come fast—far faster than anyone had realized.

CHAPTER 8

THE NEXT DECADE

TAKEOFF

Nolan County, Texas, spent the final decades of the twentieth century as a stagnant backwater. The land, just west of Abilene, in the heart of the vast, semiarid region known as West Texas, was sparsely settled. Little other than the occasional patch of cotton, a solitary cow, or a slow-swinging pump jack broke up the monotony of parched grasses and scrubby cedar and mesquite trees. Most of the terrain was flat, and this was where the towns and the railroads had been built. But a string of small, intermittent hills extends through the county and beyond, east to west, for about 150 miles. "People living near them call these hills 'the mountains' though they do not qualify in the least for the title, seldom standing more than 500 feet above the surrounding country," Abilene journalist A. C. Greene writes in his 1969 memoir.

The buffalo had once picked their way around these hills, searching for sustenance. Where the buffalo went, the buffalo hunters followed, and in 1877 the town of Sweetwater began as a small outpost to supply them. Sweetwater became the seat of Nolan County, and during the twentieth century various industries cycled through it—a Gulf Oil refinery, a telegraph center, gypsum and cement plants, and clothing manufacturers. But one by one the factories closed and the workers were laid off, and droughts transformed the nearby cotton fields into places where, in the words of one local, "a lot of people work all their life and didn't have much to show for it."

And so, in the middle of the century, even as the oil kept flowing and a mighty interstate was built through the town, the young people began their exodus to the big cities, where there was a chance to start anew. By the 1990s the oil fields were drying up, and the main thing Nolan County was known for was Sweetwater's annual rattlesnake roundup, billed, of course, as the "world's biggest."

Nolan County's story would be typical of rural America but for one thing: the renewable-energy policy signed into law by Gov. George W. Bush in June 1999. The law, one of the first and most successful of its kind in the country, transformed those barren, desolate hills where cows had once grazed unhurriedly beneath the open sky into money-making machines, generating more power from the wind than in any other county in the United States. The winds, which Greene describes as "so harsh your neck turned raw where the collar just missed the hairline by a centimeter or so," rushed up the backs of those hills and gathered speed as they plunged down the other side. These were the same winds that had given hope to Carlos Gottfried and his short-lived Hummingbird turbine factory in Sweetwater in the 1970s, the same winds that, at the end of Dorothy Scarborough's 1925 melodramatic novel *The Wind*, drove the fictional Letty Mason into the arms of the magically reappearing stranger Wirt Roddy as she shrieked, "The wind! The wind! Don't let the wind get me!" Now those terrifying winds could be harvested for a useful purpose.

Drawn by the open spaces and the winds, renewable-energy companies began to trickle in. Sweetwater's hills, according to longtime wind hand Mike Sloan, were the last sizable mesas on the way to Dallas, 200 miles to the east. The windmen could therefore count on the shortest possible distance to dispatch their power from the remote hillsides to the population centers. Better still, a transmission line already existed, running hundreds of miles from Dallas to Odessa. The line had been built to ferry power from natural gas plants, but these were decades old and coming to the end of their useful life, so there seemed to be plenty of room for a new type of power. Desperate for jobs that would replace the factories of yore, Nolan County rolled out the welcome mat, offering tax breaks. The implicit promise was that if the windmen stayed, they would find good workers and a supportive community.

In 2001, two years after Bush signed Senate Bill 7, the first of the huge wind farms in Nolan County began operating. This was a $160 million project called Trent Mesa, and its turbines were considerably taller than the Statue of Liberty, measured heel to torch. The winds it tapped were so strong they hampered construction, for workers needed a spell of calm weather to erect 200-foot towers with 110-foot blades. "That's the irony of it," Peter Main, an official with American Electric Power (AEP), remarked to the *Abilene Reporter-News*. "The site was picked after years of research because it has the right wind conditions, but then you have to contend with the wind during construction." At the time it was built, Trent Mesa, which at 150 megawatts could supply the power needs of nearly 50,000 Texas homes, was the fourth-largest wind farm in the world, and the first in the Sweetwater area.

More notable were the names of the companies that proudly announced the project, for they signaled that Texas wind power now had big, deep-pocketed backers. American Electric Power, which owned the plant, was one of the largest electricity generators in the country (in 2000 it had merged with Central and South West, the utility that had commissioned power from Southwest Mesa). Mostly, AEP focused on coal and natural gas, but now it was increasingly eager for wind. The power would be purchased for ten years by TXU, a notoriously conservative Texas utility giant, which less than a decade earlier had spurned the notion of joining the LCRA to buy power from the first big wind farm in Texas. Now TXU had become "the largest purchaser of renewable energy in Texas and the Southwest and a national leader in supporting wind power," its vice president proudly proclaimed. In these still-early days of the Texas wind industry, long-term commitments by utilities like TXU and by local electric providers serving San Antonio and Austin provided crucial assurances to wind-farm owners that their product would remain in demand for years.

The new wind farms drew back some of Sweetwater's native sons. Greg Wortham, who had grown up in Sweetwater in the 1960s, was living in New York, where he was trying to run an energy cooperative. During a visit home he noticed that next to a valley that used to be full of pump jacks, the hillside was now covered with wind turbines. It was Trent Mesa. Back in New York, a thought occurred to him: "There's talk, talk, talk up here and no action. There's action, action, action and

8.1. *A specially outfitted truck transports a wind turbine component to the Sherbino Mesa wind farm near Fort Stockton in West Texas. Courtesy Lone Star Transportation.*

no talk [in Sweetwater]." He headed back to Texas, and, with help from his father, a licensed pilot and former grocery company supervisor, he began a group that promoted the wind business and wooed more developers to Sweetwater.

Successes piled up, almost faster than Wortham had dreamed. The federal tax credit that developers had worried about in 2001 kept getting extended, providing assurances that wind power had at least a chance of competing with coal and gas and nuclear.

The turbines at Trent Mesa had been built by Enron Wind, and to this day, according to Wortham, the slanted E logo peers down from the nacelles at the top of the towers in the middle of a private ranch. But the name Enron did not appear on the press releases. The public dedication was scheduled for November 15, 2001, but it was canceled because of "inclement weather." Less than three weeks later, the red-hot company that Kenneth Lay had built into an electricity-trading powerhouse filed for bankruptcy as damning revelations about offshore accounts tumbled out. Soon, Lay would be in a Houston courtroom facing a jury, which would weigh charges of fraud and conspiracy against him.

For wind developers in Texas and across the nation, this was a profoundly demoralizing moment, harkening back to the bankruptcy of Kenetech five years earlier. Enron Wind had become one of the six largest turbine manufacturers in the world, and it had grand plans for

Texas. In 2001, even as American Electric Power began operating the Trent Mesa site, it was in negotiations to buy another wind farm called Desert Sky in Pecos County from Enron Wind. Ward Marshall, who had joined AEP after its merger with Central and South West, remembers sitting across from one Enron Wind employee, Jay Godfrey, when a lawyer arrived to announce that Enron had gone bankrupt. "I remember Jay looking up, [and he] goes, 'Hmmm—hope our credit card still works to get home,'" Marshall says.

In some ways, arguably, the collapse of Enron Wind turned out to be an unexpected blessing for the industry. Robert Bradley, Lay's former speechwriter and a critic of wind power within the company, says that Enron Wind had never turned a profit, and Lay had been trying to sell it. According to Bradley, Lay once even joked about the California turbines Enron operated being bird Cuisinarts—the term had entered the wind-power lexicon. But in 2002, when the dismantling of Enron began, the wind division was sold off to a company that had been a household name for a century, General Electric. Enron Wind was the first piece of its bankrupt parent to generate substantial cash for hungry creditors, and GE won the auction and snapped it up for some $358 million, a number later revised downward, not least because, according to a lawyer involved, no one trusted Enron's own accounting. But the takeover by General Electric meant that wind was a big boys' game now. The company begun by Thomas Edison was a global manufacturing powerhouse, and now it had decided that wind was worthy of its attention.

Meanwhile, Sweetwater kept humming along. GE arrived and took over a Coca-Cola storage facility. Another wind developer moved into the space of a business that had made deer blinds. A near-abandoned cafe began to revive. And in 2006 the world's largest wind farm, called Horse Hollow, was erected in Nolan and nearby Taylor counties, and other projects were racing ahead so fast that towers and blades lay along the county's farm roads like wood waiting to be stacked. Ranchers that had once struggled to make ends meet with cattle or hunting suddenly found a new source of income.

"You don't have to worry about whether it rains or not," said one Nolan County rancher, Johnny Ussery, who sold his cattle after a bit of soul-searching and, in 2006, decided to live off the royalties from a few dozen wind machines on his land. Plenty of other landowners did the same thing, but sometimes they kept cattle, which liked to cluster in the cool, narrow shadow of the tower, or they let their land out to hunters, as long as the hunters agreed not to drive drunk or shoot at the turbines. Rural schools bought new books and, later, iPads, with the flush of money that the wind companies had brought. (In the early days, reductions in the amount of property taxes paid by a wind farm sometimes came in exchange for big donations to local schools; in 2009 this direct-donation policy was scaled back, though wind companies could still negotiate hefty reductions on their property taxes. Either way, the mere presence of a big developer in a small town yielded hefty funds.) In 2007 the local technical college began planning for a wind-energy degree, which would be the first of its kind in Texas.

By mid-2008 Nolan County, a place slightly smaller than Rhode Island, was home to an astonishing 2,500 megawatts of wind-energy capacity, enough to power hundreds of thousands of Texas homes. The two local barbecue joints, Big Boy's and Buck's, did a booming business selling to windmen. By itself the county had more than fulfilled the goal signed into law by Governor Bush nine years earlier of doubling the amount of wind generation in the country. Nolan County produced more power from the wind than all of California, and in 2007 Wortham got elected as Sweetwater's mayor. "West Texas is the fourth largest nation in wind energy today," he told CBS that year. "There's Germany, Spain, India, and West Texas."

But Nolan County, with its wind booster–turned-mayor, was not actually the first place Texas wind developers had turned after the renewable-energy requirement passed. Enormous wind farms were going up—had gone up—some 150 miles to the Southwest, near a town in Upton County that had been quite literally established on the day in 1925 that a wildcatter there had struck oil. The wildcatter's name was George B. McCamey, and the place was named for him.

McCamey's oil boom, like others across Texas, was temporary. By the 1990s the pump jacks around McCamey were slowing, and a quiet despondency had settled over the town. When Walter Hornaday and Sam Enfield began working on the wind farm that became Southwest Mesa, McCamey was still a one-stoplight kind of place, with a Dairy Queen and a post office and a high school as well as pipelines and tanks and other necessities that grew from the oil fields. Just 1,805 people still hung on there in 2000, by the count of the U.S. Census.

In 2001 Texas lawmakers passed a resolution proclaiming that this

town, built on oil, would thereafter be known as the "wind power capital of the world." A red, white, and blue sign trumpeting the new slogan
soon went up beside the highway, and the townspeople introduced an
annual "Wind Energy Capital of Texas Cook-Off," held in September.
McCamey had great winds, even stronger and better than Sweetwater's,
and as soon as Bush signed the renewable-energy requirement in June
1999, the windmen booked flights to Midland, then hopped in rental
cars and drove the hour south to the town. "When they got Southwest
Mesa built, then those landowners saw the turbines, then it became
real," says landman Chris Crow, who watched the project with interest.
"Until then, it wasn't real." Beth Garza, then working for FPL Energy,
said the cook at a local restaurant, Benoit's, decided to branch out and
open an unprecedented second restaurant, also in McCamey.

As the boom progressed, experienced wind hands from Denmark,
Japan, Britain, and California began descending on McCamey, joining
with the local gringos and Hispanics. The influx of out-of-staters reflected the increasingly global nature of the wind business: Danish or
Japanese companies might have built the turbines, and the Brits knew
how to put them up. Randy Sowell, a Lubbock native working out there
at the time, recalls that everyone spoke English but couldn't understand
one another because of varying accents. Texas landowners would call
Sowell and tell him, "'The Brits have just called and they want something, but I can't understand what they want.'" So Sowell would call the
Brits to figure out what was going on and call the landowner back to in-
terpret. Even the local golf course featured Danish sand in its traps (and
still does), because workers at the Danish company NEG Micon had
packed the blades for the wind machines at Southwest Mesa in sand.
"I can remember the project manager guys trying to figure out what to
do with the sand," says Garza, "and somebody decided—'Well, let's just
take it down to the golf course.'"

Day after day, more turbines went up on the mesas, and wind farms
with majestic names soon spread across the region around McCamey:
King Mountain, Woodward Mountain, Desert Sky (which AEP succeeded in buying from Enron Wind in December 2001), Indian Mesa.
They went up on the flat-topped hills, many of which already had roads
running across them because oil had been pumped in the area for generations, and they went up quickly because the wind industry feared
that the production tax credit that guaranteed lower-cost wind power
for ten years would expire at the end of 2001. King Mountain, at 280
megawatts, remained the largest project in the world for years.

8.2. *The King Mountain wind farm, near McCamey and completed in 2001, was for years the largest wind farm in the world. Courtesy Walt Hornaday/Cielo Wind Power.*

The landowners were curious about the new-style machines that cranked out money. Mr. Woodward, the nonagenarian owner of a huge tract, was said to prowl around the top of his mesa to make sure the men putting up the 160-megawatt Woodward Mountain wind farm were doing their jobs right. He drove a Jeep, it was said, because his family did not want him to ride a horse anymore.

In their rush to farm the new energy on McCamey's mesas, which reached 2,000 to 3,000 feet above sea level, developers overlooked one critical detail. McCamey lay more than 300 miles west of Austin and almost 400 miles west of Dallas. Only three transmission lines, and relatively small ones at that, led out of the area.

The result was a power bottleneck. When the wind was blowing at twenty or thirty or forty miles per hour atop the Upton County mesas, as it often did, as much as 40 or 50 percent of the power generated by the new wind farms could not go anywhere. Imagine enticing lots of people to work downtown, says Beth Garza, then of FPL, but failing to build more roads into and out of the place. While wind farms could be built in a matter of months, transmission lines could take years and required a painstaking public process to site.

And so, on the windiest days in the early 2000s, many of the turbines around McCamey would simply cease spinning. The mesas were covered with large metal and fiberglass machines, the latest technology from Danish or Japanese companies, but operators turned them off and they produced no power. Wind developers may simply not have realized just how many turbines were going up on the West Texas hills until they ran into rivals at the Midland Hilton or saw their white trucks on the roadways. (When wind developers filed applications to "interconnect" with the electric grid, they kept them secret so that competitors did not find out about their plans.) And so as the turbines, like the oil fields before them, began to produce less energy, the developers got in their trucks and headed away from McCamey, northeast, toward Sweetwater and Nolan County, where the winds were a bit less powerful but at least there was a big wire, with some spare capacity, that ran all the way to Dallas.

By 2006, however, Sweetwater was facing similar problems. The giant turbines, whose operators could use modern electronic controls and ever-improving forecasting equipment, produced too much power for the single wire running to Dallas to handle. When the wind was blowing across the mesas of Sweetwater, as it often did, some turbines at the mighty wind farms like Roscoe and Buffalo Gap had to shut down. Drive down a road at that time, says Greg Wortham, and "you look up and there's hundreds of turbines off on a windy day." So urgent was the situation that in 2009 the energy giant NextEra Energy Resources, the former FPL Energy, which operated the wind farm known as Horse Hollow, one of the biggest of them all, used its own money to build a 214-mile private transmission line to ferry its power to a place that could use it, bypassing the usual protracted public process for erecting transmission lines.

But the wind companies remained confident about the state's commitment to wind power: if politicians in the oil and gas state wanted to burnish their environmental bona fides and aid struggling rural areas, the wind developers reasoned, then they would build power lines to alleviate the problems. And Texas officials for the most part obliged. Rick Perry, a cowboy-booted, free-market West Texan who had inherited the Texas governorship in 2000 as George W. Bush headed to the White House, had shown himself to be, if anything, more of a wind cheerleader than Bush. Perry traveled to West Texas for occasional wind farm ribbon cuttings, and in 2005, the governor put wind on the agenda during a special session of the Texas Legislature. When lawmakers produced a bill that more than doubled Bush's renewable-energy mandate, Perry signed it, without demanding any deregulation-style quid pro quo. Now the state was required to get 5,880 megawatts of renewables capacity installed by 2015 and would aim to have 10,000 megawatts by 2025. It wasn't a huge leap—less than 10 percent of the state's peak demand, Perry's spokesman would say much later—but Texas, once again, ended up getting it done even faster. Within six years electric companies had exceeded the 2025 goal. "[Rick Perry] has been a stalwart in defense of wind energy in this state—no question about it," Paul Sadler, then executive director of the Wind Coalition, told the *Texas Tribune* in 2011, the year Texas topped 10,000 megawatts.

The bill Perry signed in 2005 did something else, too: it required Texas to start mapping out a web of transmission lines that would solve the perennial problem of too many turbines and not enough wires. The next year Perry traveled to Dallas, where, surrounded by renewable-energy executives, he pledged that investments of $10 billion would flow into Texas wind, as utilities built power lines to complement private companies' construction of wind farms. It was this pledge that made Sweetwater's wind developers confident that if they kept on building, the wires would soon come. Perry himself had this to say during an October 2006 visit to Amarillo, where he spoke at a wind rally during the height of his reelection campaign: "Wind energy is good for the environment, good for the economy and good for West Texas, because there is an ample supply. Last month I announced a $10 billion wind energy investment, and I am going to work hard to bring as many of those wind turbines as possible to Amarillo and the Panhandle. I might even put up a wind turbine or two at the campaign headquarters of my opponents to capture all that hot air."

By 2008 Texas regulators appointed by the governor had a plan. The state would order transmission companies to build more than 2,300 miles of new lines—enough to crisscross the entire state several times at least—to aid wind power. The web of lines, called CREZ, for Competitive Renewable Energy Zones, would be centered in West Texas. They would solve the problems in Sweetwater as well as, finally, in McCamey, and they would also reach north into the Texas Panhandle, the windiest part of Texas but one of the remotest. It would more than double the amount of wind energy already in the state, at a projected cost of nearly $5 billion. Texas could build these lines more easily than other states because it had its own electric grid, unlike other Lower 48 states, which were on the eastern or western grids. That meant utilities were building the power lines all within a single state, so they did not have to get approval from multiple states or the federal government. Indeed, the Texas electric grid, which covers about 75 percent of the land area in Texas (the Panhandle, El Paso, and parts of East Texas are on the other U.S. grids for obscure historical reasons), had evolved on its own precisely in order to avoid federal oversight, which generally kicks in when power flows between states.

The billions of dollars for transmission, which would come out of Texas ratepayers' pockets, was an astonishing commitment from a state that resents taxes—or anything resembling taxes. It was an irony of policy making in a libertarian-minded state that these lines would be paid for by a socialized fee, payable by all Texans whether they bought wind power or not. It was also remarkable in a state that has deep respect for private property rights that companies building the lines could seize the land they needed through eminent domain if the landowners stood firm and refused the one-time payments of the utilities.

Some landowners in Texas did stand firm. Far more than the wind farms themselves, the transmission lines provoked opposition. They are the ugly stepchild of the electricity business—unsightly, unglamorous, and hard to build. Whereas wind turbines have a certain elegance, spinning gently and easily in the sunlight, transmission lines are simply wires stretching across the horizon, ferrying electrons from the point where they are created to the homes and businesses where they are used. Seen from the air, they cut through the landscape like a long, thin scar. From the ground, the huge steel towers supporting the lines can be seen for miles.

The epicenter of the opposition turned out to be in the rolling and beautiful land of the Hill Country in Central Texas, which lies between the mesas of West Texas and the electricity-hungry population centers of the central and eastern parts of the state. This was the land where Lyndon Johnson was born in a tiny house, and where he had worked hard to bring rural electrification to his constituency. Back then the Hill Country was poor. But by the twenty-first century, because of its juniper-covered hills and rivers and peaceful vistas, the area had become one of the most desirable spots in Texas. Wealthy Texans saved up to buy Hill Country ranches, where they could retire and gaze out at pecan trees and peach orchards and even, as a lure for the tourists from Austin or San Antonio, enjoy a few wineries.

And so around 2009, an almighty fight began, in the hearing rooms of local judges and of the Texas Public Utility Commission, which oversaw the process of selecting routes. Landowners hired Austin lawyers to lobby the commission to reroute lines, which would be built in the Hill Country by the Lower Colorado River Authority, the very same utility that had adventurously backed the first wind farm in Texas more than a decade earlier. In a twist, some of the fiercest opponents were environmentalists, who argued that the transmission lines meant to deliver green power would actually harm the countryside. Bill Neiman, owner of the Native American Seed Farm, which grows native plant seeds along a river a couple of hours west of Austin, told the *Austin American-Statesman* that his property and livelihood would be ruined by wind transmission lines if they went through his land or even near it. He built a scaled-down model of a transmission tower and, in protest, toted it in a flatbed truck to public meetings.

As the process wore on, novel arguments emerged. People cited Indian burial mounds, historic stone walls, unusual landforms, endangered songbirds, a large bat colony, and anything else they could think of as a reason that the lines should not cross their land. The tiny town of Clifton, an artists' colony northwest of Waco, argued that the lines

8.3. *To protest a proposal to string transmission lines across the Hill Country to bring West Texas wind power to the state's metropolises, Bill Neiman built a scaled-down model of a transmission tower and toted it in a flatbed truck to public meetings. Courtesy Bill Neiman.*

would thwart its ability to attract painters and sculptors. If a line was built skirting the city, "Why would these artists come?" Fred Volcansek, the mayor, asked Texas regulators in 2010, as a hundred residents of the town looked on. Volcansek lost. At the end of the day, the Public Utility Commission was unmoved by the claims about art and landscape and chose a route near the town.

Hill Country landowners were luckier: three big lines had been planned for that area, but they succeeded in getting one canceled and slashing the length of a second. Somewhat extraordinarily, the three public utility commissioners, faced with such vigorous opposition, decided that despite losing those lines, the grid could achieve essentially the same result by simply upgrading existing substations and wires, saving over $140 million in the process. Some land in the Panhandle near beautiful Palo Duro Canyon State Park, another flashpoint, was also spared.

The process of deciding the routes is now complete, and landowners along them are settling in for a lifetime of staring at high-voltage lines.

The projected costs of the entire system have risen astronomically—from $4.9 billion in 2008, when the lines were announced, to $6.8 billion in 2011, or about $270 for every man, woman, and child in Texas. The increase came after the state revised its calculations to reflect that the lines would not march straight across the land from Point A to Point B, but rather would often hug property lines and roads to minimize disruption to landowners. Yet, led by one of the most conservative governors in history, Texas has pushed forward undeterred despite the higher costs. The lines should be finished by the end of 2013, clearing the way, in theory, for another wind farm boom, including in the remote hinterlands where Texas wind was born.

CHAPTER 9

THE FUTURE

One afternoon in May 2008, T. Boone Pickens—eighty-year-old oilman, corporate raider, and the author of a new book, *The First Billion Is the Hardest*—pulled into the Panhandle town of Pampa. He was headed for the M. K. Brown Memorial Civic Auditorium, and when he got there, about 300 people—ranchers, kids, grandparents, fire fighters—were packed in, waiting. Pickens had a proposition for them: he wanted to erect about 2,700 wind turbines in the flatlands, generating enough electricity to power 1.3 million homes. It would be, he said, the largest wind-power project the world had ever seen, and it would bring jobs, money for the school district, and a cut of electricity profits for landowners who hosted the machines on their property.

Pickens had gotten interested in wind power as the price of one of the energy sources he knew best, natural gas, had soared. These were the last days before the fracking boom, when drillers probing under the earth would be able to extract more natural gas than anyone thought possible. Those supplies were hidden in tight shale formations, but they could be blasted apart with a high-powered combination of water, sand, and chemicals. Soon, companies drilling in Texas and Pennsylvania and Ohio would announce gas discoveries big enough to last the country 100 years, even as controversy erupted over the drilling method.

But at the time of Pickens's appearance in Pampa, natural gas prices were nearly at an all-time high. They had ridden a roller-coaster since Hurricane Katrina had disrupted production in the Gulf of Mexico three years earlier. The high prices kept wind power—aided, as always, by the federal production tax credit—competitive, because in Texas natural gas at that time accounted for two-fifths of the fuel on the electric grid, and its price essentially dictated the overall price of power. Companies eagerly bought wind power, not necessarily because it was green but to guard against long-term fluctuations in the price of fossil fuel–based electricity, and plenty of ordinary Texans did the same. In 2006, when the price of wind power dipped below that of conventional electricity for a time, Austin's electric utility, Austin Energy, had so many residents hoping to buy the cheap, green wind power that it was forced to hold a lottery for the privilege.

It was in these post-Katrina years of high natural gas prices, coupled with signs of a commitment by Texas officials to build transmission lines, that wind began to attract serious interest as far away as Wall Street. BP and Shell ramped up their wind investments, as did the likes of Wells Fargo and J. P. Morgan Chase. Even Goldman Sachs, the mighty New York investment house, bought a Houston wind firm, Zilkha Renewable Energy, in 2005. Wind, Zilkha explained, was the "fastest growing segment of the energy market," and that was why Goldman wanted a piece. Pickens did, too, and he told that to the audience in Pampa.

"How many of you have heard me say Pampa will be the wind capital of the world?" asked Pickens, to a sea of nods. "You like that, don't you?"

Pickens owned a ranch not far from Pampa and struck the pose of a benevolent uncle. "Friends, neighbors, and partners, we'll do a first-class job," he continued. "I'd like to think I have credibility in this area."

Pickens turned to the whiteboard as he explained what he had taken to calling the "Pickens Plan," a patriotic case that the country needed to reduce its dependence on foreign oil and, as a substitute, use more American-drilled natural gas for cars and more American-made wind power for electricity.

The U.S. Department of Energy would give credence to the notion of large-scale wind that summer, when it put out a study saying that wind could theoretically produce 20 percent of the country's electricity by the year 2030. Pickens wanted to make that happen, and more.

A woman asked whether the giant windmills would make noise.

"I've been a quail hunter since I was twelve, so my hearing isn't worth a hoot," Pickens told her. "If you're getting royalties from it, it might have a real pleasant sound."

But the turbines have made no sound at all. Despite Pickens's grand pronouncements, the Pampa wind project never got built, and in the corridors of wind conferences the mention of Pickens's name soon brought snorts of irritation. A few months after his appearance at the Pampa auditorium, the price of natural gas began to plunge as the extent of the enormous new shale supplies became clear. As the price of gas fell it pulled the price of all forms of electricity down with it, and wind became less competitive. "When natural gas is $4.50 [per thousand cubic feet], it's hard to finance a wind deal," Pickens told the *Texas Tribune* in 2010, the same year he gave up the last of the leases on the Pampa land. "Natural gas has got to be $6." The credit crisis had dealt a final blow, causing financing options for risky Texas megaprojects to disappear rapidly and leaving Pickens scrambling to figure out what to do with hundreds of turbines he had ordered from General Electric. In 2012 Pickens announced plans for a much smaller wind farm south of Lubbock, at 377 megawatts, less than one-tenth the size of his vision for Pampa.

Pickens was not the only one with dashed dreams. Along the Gulf Coast, an aging entrepreneur named Herman Schellstede, whose father in 1947 had helped construct the first offshore oil rig in the Gulf of Mexico, had begun thinking about wind after decades of designing offshore oil and gas platforms. A Sierra Club friend who owned a Cadillac dealership, as Louisiana environmentalists are wont to do, had told him that wind was the next big thing, and Schellstede, who knew how wind could batter isolated rigs in the water, took him seriously. And so in 2004 the man whom the *Financial Times* would later describe as "an oil man's oil man" got into the business of offshore wind power and searched around at banks and oil companies for over $300 million to build a sixty-two-turbine farm nine miles off Galveston, in fifty-foot water. It was a lot of money—offshore projects cost considerably more than onshore ones—but the winds, he knew, were far stronger and steadier than the ones in West Texas. Near to shore, the Gulf is relatively shallow, which "gives us a big advantage over the New England boys," says Schellstede. In a convenient quirk, Texas waters extend up to ten miles offshore, considerably farther than most states, due to historical reasons relating to how Texas joined the union. This means that developers like Schellstede had plenty of room to plant turbines without hitting federal waters and triggering a cascade of new rules. Also, an enormous offshore service industry already existed in Houston, Galveston, and Corpus Christi.

In 2007 Schellstede's company, Wind Energy Systems Technology, put up a meteorological tower to measure the wind. A year later Hurricane Ike struck. "As the hurricane approached us, the winds went up to 110 miles per hour," says Schellstede. "Then they went down to zero as the eye passed over." Then the other wall came, with sustained winds of 140 miles per hour and gusts that may have reached 200 miles per hour. "The gusts are the really ferocious items that really tear things apart," he says. Somehow the tower survived, and Schellstede emerged with plenty of new details for his "wind profile" of the site. Later, one local asked if Schellstede would put a restaurant on top of his tower so diners could see the turbines when they were up and spinning. Otherwise, "You could barely see them if you were standing having a margarita in Galveston," Schellstede said.

Schellstede had hoped to get the wind farm up and running by 2010, and he secured five leases from the Texas General Land Office, which controls the state's offshore lands. "Oil and gas is a diminishing resource," was a favorite saying of Jerry Patterson, the land commissioner who took great pride in arguing that Texas could build wind farms in the Gulf faster than Massachusetts's Cape Wind project or other wind farms planned off the East Coast. Patterson put it this way to the magazine *Fast Company* in 2008: "We don't have Ted Kennedy in Texas, so we don't have anybody with the hypocrisy of 'I'm in favor of green power. Oh, but you're going to put it here off my house? No, no, no we don't want that.' We have people who are realists. I don't even get into the debate about global warming. It's an argument that has no justification because we need to be doing the same things whether global warming is man-made or not. We're running out of hydrocarbons." Asked about birds—a big concern for environmentalists, because the Texas Gulf Coast lies in the pathway of major migrations—Patterson told the reporter, "I talked to the Audubon Society and told them, 'Don't worry about this, after several generations we'll have smarter birds.' They did not think that was funny. The other thing I told them was wind farms in the Gulf of Mexico would be the first line of defense against avian flu. These people have no sense of humor. You can't break the ice with them."

So far Schellstede's plans, like Pickens's, have fallen short. In 2008 the credit crisis struck, wiping out potential investors like Lehman Brothers and Wachovia. That event "put us on our knees," said Schellstede. The plunge in natural gas prices also made it harder for an offshore project to compete against more mainstream technologies, as have the uncertainties related to federal policies on renewable energy. Schellstede remains a perpetual optimist, predicting in early 2011 that he would have a turbine in the waters off Galveston by the end of the year. Also that year, an Austin-based company, Baryonyx, proposed erecting 200 offshore wind turbines between Corpus Christi and Brownsville. Neither plan has yet been realized. In early 2012, unfazed, Schellstede was still determined to put up that solitary turbine, and he spoke eagerly about the day that five offshore wind farms built by his company would line the Texas coast.

The future of Texas wind lies, for the moment, in the very places that Pickens and Schellstede could not make it work. The biggest hopes are the Panhandle and the coast, though mostly onshore rather than off. Pickens had picked Pampa not only because his folksy ways would resonate with the landowners, but also mostly because the area sported some of the state's most consistent winds. This was the same place, after all, where nearly three decades earlier hometown boy Michael Osborne, inspired by watching a glider soar and swoop during his childhood, had plugged five spindly turbines into the grid in 1981 and pronounced it the first wind farm in Texas. Yet the Panhandle's far-flung isolation also meant that even as the West Texas sites with lesser winds like Sweetwater and McCamey got developed, the land of the High Plains remained much as it had always been, with nothing taller than the occasional farmhouse or grain elevator. The *Texas Tribune*, in an article titled "Remoteness a Hurdle in Harvesting Panhandle's Winds," put it this way: "The Panhandle's challenge is its remoteness. Few transmission lines exist to carry power to big cities that need it. (A similar problem plagued wildcatters in the 1920s, when oil was discovered in the Panhandle and got excavated faster than pipelines or railroads could be built.) Much of the Panhandle isn't even on Texas' electric grid, called the Electric Reliability Council of Texas, or ERCOT; instead, it's part of a grid called the Eastern Interconnection, which runs all the way to Maine."

Yet by 2010, the year that article was published, the Panhandle was shaping up to be one of the next great sites for the Texas wind rush. The buildout of transmission lines planned by the state would reach deep into the Panhandle prairies for the first time, going even beyond where the traditional boundaries of the Texas electric grid ended north of Midland, to draw the wind back to Texas metropolises hundreds of miles away. Records showed a healthy queue of prospective projects, including more than 2,000 megawatts of new wind power planned for Gray County, the old stomping grounds of Michael Osborne and Woody Guthrie. Several farms were proposed for nearby Briscoe County, with its steep cliffs of caprock, where winds pick up as they swoosh over them. Silverton, the county seat, is "probably the epicenter of the biggest development you'll see," predicted wind hand Mike Sloan in 2011. He noted that the area had already seen a big land rush by developers. One Briscoe County wind farm had been proposed at a whopping 2,940 megawatts of capacity, which would be four times the size of the largest wind farm in the world (located, of course, near Sweetwater), and one and a half times the size of the entire 1999 wind requirement signed into law by George W. Bush. Put another way, it would offer more than 23,000 times the capacity of Osborne's 1981 effort in Pampa.

The other up-and-coming wind site is the Gulf Coast, where breezes blow in from the sea. Although less powerful than the legendary West Texas gales, the Gulf winds are strong enough to nearly flattop the oak trees clinging to life atop the sand dunes. They are also better matched to the needs of the Texas electric grid, because air conditioners kick into high gear in the late summer afternoons, just as West Texas breezes tend to stall but Gulf breezes pick up. The coast has easy access to transmission lines, unlike in West Texas, because a dense network of wires already exists to serve the millions of people who choose to live near the water. Even as construction began on the expensive transmission lines to West Texas, wind developers discovered that there was plenty of potential in a place that needed no new lines.

The coastal turbines haven't gone up without a fight: bird lovers, fearful that migrating flocks could get chopped up, found an unusual ally in the 825,000-acre King Ranch, a place described by *Texas Monthly* as being "the first ranch in Texas, the cornerstone of the cattle business in the West," and "instinctively hostile to outsiders." When officials at the neighboring Kenedy Ranch became interested in hosting wind farms, the King Ranch directed some of its famous hostility toward them, teaming up with the environmentalists to sue to stop them. "I think it's a tragedy for the state and the coast," complained Jack Hunt, the King Ranch's former chief executive, who points out that the navy's air station nearby also has concerns about radar interference.

The windmen had a sharp response. "It's impossible to build a wind farm anywhere in the U.S. without someone thinking it's a bad idea," Ward Marshall, then helping the developer, Babcock & Brown, prepare sites on the Kenedy Ranch, told the *Austin-American Statesman* in 2008. "The first thing opponents will grab on to is an environmental issue: With coal plants, they hang their hat on air quality; with nuclear plants, they worry about a Chernobyl meltdown; and with wind farms, it's a bird problem. That's our albatross."

The suit brought by the green groups and the King Ranch failed—Texans, after all, have little say about what their neighbors do with their land—and wind developers are moving ahead, planting turbines on big ranches and coastal cotton fields, where workers have been known to spot alligators in drainage ditches. "The short term for coastal wind is great," Patrick Woodson, an old wind hand who by then had joined the North American arm of a European utility called E.On, told the *Texas Tribune* in 2011. "There will be a number of prime sites that get built out in the next two to four years." By early 2011 South Texas accounted for roughly one-ninth of the state's total wind-power capacity, and during a summer of record-breaking heat that same year, the coastal turbines were lauded by electric grid officials for supplying late-afternoon power when the system was badly strained.

CHAPTER 10

THE LESSONS
OF TEXAS WIND

Around 1980, shortly after founding the Alternative Energy Institute, Vaughn Nelson paid a visit to Panhandle farmer Joseph Spinhirne. In the 1940s Spinhirne had stood fast against rural electrification. Back then he bought a Jacobs wind charger, and when Nelson visited, the unit was still producing power for the air-conditioning and other appliances, while a nearby Aermotor windmill pumped water.

"We owe a great debt to people like Spinhirne," Nelson mused to Jon Naar, who accompanied him. "They taught us to stay with natural energy and to use it self-reliantly. The wheel has turned full cycle. Forty years ago they may have been considered out-of-date, but in the 1980s they are showing us the way ahead."

Just so, the wind pioneers of the 1970s were showing the way ahead for modern Texas, though they did not fully realize it at the time. Decades of shimmying up towers to gather wind-speed data and running tests that sometimes ended in disaster helped lay the groundwork for the great Texas wind rush that accelerated after 1999. "When the wind farm boom took place in California from '82 to '85, I never thought I'd live to see the day that we would have wind farms in Texas," Nelson told his university's magazine in 2010. That conservative, oil-loving Texas would put in place the incentives needed for a similar wind spree seemed inconceivable.

Yet Texas beat out California and every other state in the country, and the lesson is one of persistence. Men like Osborne and Nelson knew how the government had shelved Percy Thomas's aerogenerator sketches in the late 1940s, killing off interest in wind for decades, and they were determined to avoid losing the thread again. So even during the dark days of the 1980s they never stopped peddling their ideas to anyone who would listen. Bit by bit, politicians in the enormous, windy state swung around. "As someone who grew up in West Texas, I can vouch firsthand for the fact that there is great wind potential out here," Gov. Rick Perry, born and raised in Paint Creek, said in 2002 at the dedication of the Desert Sky wind farm in Pecos County. Plenty of other elected officials could say the same.

By 2011, "powered by Texas wind" had become a badge of state pride, as politicians from Perry on down learned to whoop gleefully about the oil and gas state as an improbable trendsetter for the national "green" movement. That year wind accounted for 8.5 percent of the power on the Texas grid, far above the national average of 3 percent. Local utilities in green-minded cities like Austin and San Antonio have led the way, but companies as random as HostGator, a Houston-based Web-hosting company, buy credits from Texas wind farms to offset the emissions associated with their servers' energy use. Even the electricity at the 2011 Super Bowl at Cowboys Stadium in Dallas was offset by wind energy generated at Sweetwater. That included the power for the once-largest-in-the-world flat-screen television, the Black-Eyed Peas halftime show, the locker-room lights, and the refrigeration for all the Coors Light and Coke that kept fans hydrated.

The new industry has led to new jobs—twenty-first-century versions of the oil roughnecks. In the summer of 2008, with only $700 to his name and his dream of sustainable farming looking increasingly unlikely, a short, weedy, thirty-four-year-old Austinite named Aron Brackeen cast around for his next job. He had long been interested in alternative energy, and soon after a phone call to an admissions officer at Texas State Technical College in Sweetwater, he struck out for West Texas. By that November he found himself atop a towering wind turbine as part of a "climb test" designed to make sure the students had the intestinal fortitude for high-up work. "I was willing to follow the wind," said Brackeen, who later got full-time work at a Sweetwater wind farm. By 2010, nearly 10,000 jobs existed in Texas wind, according to a report by the Perryman Group, an influential Texas analysis firm.

A wind-power cottage industry has sprung up in Texas, creating all sorts of positions that are the envy of other states. Houston, eager for all things energy, has become something of a wind-power capital, as developers including Iberdrola Renewables, BP Wind, Pattern Energy, and Horizon Wind Energy expand their offices there. Vestas, the giant Danish turbine maker, opened a research and development arm in Houston, too, though in 2012 the company began the process of closing the office down as a cost-cutting measure. (Jobs in wind manufacturing in Texas have generally been sparser than in other states, owing to policy makers' reluctance to offer blade or tower manufacturers special incentives.) A couple of blade repair companies have set up shop in the Texas hinterlands. In the fall of 2011 Texas Tech University, which had already built a wind tunnel with a tornado vortex simulator, began offering a bachelor of science degree in wind energy. "As the renewable energy field continues its rapid growth and development, the wind energy job market will continue to broaden," announced engineering professor and degree program director Andy Swift.

In perhaps the sharpest signal that the wind business has shifted from a handful of fly-by-night tinkerers to an industry with money at stake, the Wind Coalition, the regional partner of the American Wind Energy Association, spent as much as $635,000 on lobbying in Texas in 2011, according to state records, including as much as $400,000 to pay its executive director, former state lawmaker Paul Sadler, who in 2012 reentered the political arena when he ran for the U.S. Senate as a Democrat. In a state where money talks, wind has a voice, even if it is still not nearly as loud as that of the big oil and gas spenders.

Lacking the sort of opportunity that came with the grand bargain that established the renewable-energy mandate in the legislature, the solar industry has been left to envy the wind industry. "The region obviously made wind work years ago, and we can do the same with solar," Greg Wortham, the mayor of Sweetwater and its chief wind booster, said at a summit in San Angelo in 2012. "This area is primed for it. Why shouldn't our area get the big [solar projects] rather than saying, 'We don't know anything about it, go to Arizona?'"

But while solar companies have pressed Texas lawmakers for a "non-

10.1. *A wind turbine rises on a hill looking over Sterling City in December 2008. The generators have become a common sight in West Texas. Photo by Jay Godwin, Austin American-Statesman.*

wind" mandate, big power providers are unwilling to play ball the way they did in 1999, when deregulation was at stake. Generating electricity from the sun costs more than using wind turbines despite a plunge in solar-panel prices, and manufacturers and businesses have argued that a nonwind mandate would drive up their rates. In 2011 Governor Perry threatened to veto a 500-megawatt nonwind proposal, which then failed to get out of a house committee. Another proposal to allow monthly fees to be levied on ratepayers to fund solar rebates that Texans could tap was also nixed.

If the solar lobby is struggling, part of the reason is the rightward shift in the political climate in the decade since Senate Bill 7. The grand bargain took place a decade before the Tea Party, and back then mandates, while not popular, still fell within the margin of acceptability. Under the new politics, wind, too, is falling under suspicion. To some degree, Texas wind has thrived because of a confluence of three government programs, the type that state leaders in the Tea Party era love to rail against: a mandate; a giant transmission buildout carried out by still-regulated utilities whose costs are distributed—"socialized"—among all Texas ratepayers; and the federal production tax credit, which has been aiding wind since 1992. Tellingly, though the credit got extended in January 2013, Texas's representatives in Washington had not clamored loudly for its renewal, unlike their colleagues in wind-friendly states like Iowa or Colorado. Governor Perry, running for president in 2011, called for an end to all federal energy subsidies, which would have included the wind credit, and he did not join more than twenty governors, including fellow Republicans from Oklahoma and Kansas, in writing President Obama to urge its extension.

Indeed, as wind has grown, Texas politicians have found themselves forced to balance a range of competing interests. These include, of course, the aviation and birding concerns along the coast and the howls against transmission lines from outposts like the Hill Country. Plenty of landowners in the Hill Country are dead-set against turbines, too, and a smattering of other opposition groups have popped up around the state, with Web sites like www.stopwindturbines.com. "Never in the history of the world have we put up 400-foot-tall blinking behemoths everywhere," West Texas landowner Dale Rankin, who sued to stop the march of wind turbines over hillsides near his Abilene-area home, told the *Austin-American Statesman* in 2007. Living close to hundreds of turbines, Rankin said, is like being "next to an airport where the jets are running their engines all the time." But in a state that welcomes development, Rankin's lawsuit, the first significant one of its kind in Texas, failed in 2006.

Perhaps the more enduring clash is between newly ascendant wind and natural gas. In a March 2, 2010, article titled "Natural Gas Tilts at Windmills in Power Feud," the *Wall Street Journal*'s Russell Gold reports that "a bitter fuel fight in Texas points to a different future: one in which gas and wind are foes." The tension has arisen because wind's ascent in Texas has come at the expense of gas. In 2005 wind barely registered on the state's power grid totals, at just over 1 percent, while gas supplied 46 percent of the grid's needs. By 2011, even after the shale-fracking boom got under way and natural gas supplies swelled, gas had slipped to 40 percent, and wind was at 8.5 percent, more than half the share of nuclear. Wind mostly has displaced gas rather than the other big player, coal, because gas has been more expensive than coal (although by 2012 coal, too, was suffering). Adding to the indignity, natural gas plants must be ready to crank up their power when the wind dies down, because they are easier than coal plants to turn on and off. And as they take over a greater share of the grid, wind farms help drive down overall electricity prices in Texas, because wind operators do not need to put anything into their plants to produce electricity and therefore can offer their power to the grid essentially free, unlike coal or gas. Indeed, on some windy nights when the blades are turning but electricity use is low, or when the grid is congested with lots of different plants offering power, parts of West Texas see "negative pricing," in which wind plants pay a modest amount to offload their power (the federal production tax credit ensures it's still worthwhile for them to do this).

The gas industry complained to the *Wall Street Journal* that wind did not have to pay for backup power, to make up for when it slowed down. "Wind shouldn't cause problems that other people have to fix," a representative from NRG Energy told Gold. The wind lobby fired back. "This is insidious," argued Woodson of E.ON. "Our competitors can essentially impose additional costs on us, and that really troubles us. This isn't death by a thousand cuts—it's death by a thousand grenades."

The fight goes to the heart of the biggest problem clouding the future of wind, in Texas and elsewhere: that it can be counted on only intermittently, unlike fossil fuel plants, which can operate around the clock. When the wind does blow, it's not necessarily at the most useful times, which makes Texas grid operators, even armed with constantly improving forecasting tools, wonder how much more wind they can handle without unbalancing their system.

Sometimes things work out. In February 2011 wind farms all across Texas got praise for pumping large amounts of power into the electric grid during a deep freeze that managed to knock out a quarter of the state's coal and gas power-plant units and caused rolling blackouts throughout the grid, even though a few turbines did go offline due to dangerously high winds and hydraulic-equipment freezes. But three years earlier, when a cold front moved through Texas and the winds died, the Texas grid operator, ERCOT, barely averted blackouts. (The wind industry says the cold front was predicted and the grid should have been prepared for it.) And on at least one scorching August afternoon in 2011, wind farms produced only about 1.3 percent of the grid's electricty, prompting the *National Review* to run a piece headlined "Texas Wind Energy Fails Again," by wind skeptic Robert Bryce.

To prepare for times when the wind stops blowing, Texas is hoping to jump-start energy-storage projects, from conventional batteries (the technology for which has changed remarkably little since Marcellus Jacobs's time) to caverns of compressed air. A few modern-day dreamers are hoping to build projects to export the bountiful energy from Panhandle turbines to other states that are hungry for Texas wind. But all of these are expensive—and even if they were to happen, it's clear that no matter how many turbines go up in West Texas, they won't oust coal, nuclear, or natural gas, at least not in our lifetimes. Wind may be clean and green, but even if the price of natural gas goes back up and the production tax credit keeps getting extended, wind must fall in line behind more reliable conventional sources, especially if legislation to cap greenhouse gases and address climate change remains in the realm of environmental fantasy. One of the best arguments for wind in a state like Texas, where the weather is as boom-and-bust as the oil industry, is that the turbines use no water; in the terrible drought of 2011, when cotton crops withered and huge industrial plants anxiously watched rivers dry up, the big turbines of West Texas had nothing to fear.

Because it is free of emissions and needs no fuel and no water, wind has become one of the fastest-growing electricity sources across the nation. In recent years wind turbines have popped up across the Great Plains and as far west as the rolling hills alongside the Columbia River in Oregon. To serve those markets, companies have built factories in the hinterlands, churning out blades and towers that get moved to their final destination by oversized trucks or railcars (in Texas the turbines often arrive from abroad through the port of Houston). Already the equipment is so large that travelers on lonely highways, watching wind blade shipments creep along, commonly mistake them for airplane wings. The machines have been getting larger, because higher towers reach better winds and a broader sweep of the blades produces more energy, but manufacturers are beginning to worry that the equipment is too big to squeeze under highway underpasses or to safely navigate sharp turns on country roads.

For wind developers those are good problems to have, and without Texas, the scale of wind would not have risen so dramatically. In a practical sense, the Texas wind boom has served as a model for other states hoping to kick off a similar spree. When Texas passed its renewable energy requirement in 1999, it was one of the first in the country, and the first to produce significant results. "We proved that you could put a relatively large amount of wind into the system without driving up costs dramatically for consumers," says Jim Marston of the Environmental Defense Fund. Now, more than half the states have similar renewable-energy requirements, though a federal one, eternally sought by the wind companies, remains elusive.

Texas has also exported what it has learned. As other states ventured into renewable energy, experienced Texas wind hands fanned out around the nation, looking for the next great site on the prairies or hilltops. Longtime Texas developers, like Patrick Woodson and Andy Bowman, have ventured as far afield as Oregon to work on wind. Dale Osborn left Kenetech in late 1995 but kept working in the field. When turbines sprouted like bluebonnets on the 2,000-foot mesas of West Texas, it helped wind-energy manufacturers in the United States and Europe get their footing economically, so that they could scale up and improve their production.

More basically, the Texas wind rush has built faith in wind. It has grabbed people's attention because, well, it's Texas. Whereas California's buildout in the 1980s was a predictable move by an environmentally over-the-top state—"just this goofy Governor Moonbeam, we all know California's crazy so we can just ignore it and not pay any attention"-type situation says longtime wind hand Tom Gray—Texas had "a whole different mystique." Nobody ever accuses Texans of having anything to do with moonbeams. Gray also remembers that the deliberative polling results out of Texas caught the eye of the American Wind Energy Association, where he was working. "We certainly trumpeted it far and wide—whoa, all these consumers in Texas want renewable energy and energy efficiency, and not only do they want them, they want them by enormous margins," he says.

And so it was left to California, which got passed by Iowa in 2009 for the number-two spot in the national wind rankings, to concede gracefully. "Texas is for us now sort of an inspiration, weird as that may seem," says V. John White, a leading California renewables advocate, who parachuted briefly into Austin during 1999 to offer advice while Texas was trying to figure out deregulation and wind. Years later Jerry Brown, in his second coming as governor, decided to take a Texas-style approach to renewable energy by trying to slash through the pile of red tape and regulations that have stymied wind development in California since he was in office the first time.

Texans, for their part, stand ready to claim the credit. "Whatever stereotypes people might have," says Marston, "if Texas can do it, hell, anyone can do it."

POSTSCRIPT

The old-timers, when they speak at wind industry conferences where they once scraped for money, shake their heads and marvel at how everything has changed. As of 2012 plenty of them remain in the game. VAUGHN NELSON retired in 2010 from the Alternative Energy Institute, but he's still writing about wind, and the AEI is still crunching numbers and testing turbines in Canyon. It recently had to move its Canyon facility once more after creating trouble yet again. "We put up a turbine that was a real crappy turbine," says AEI assistant director KENNETH STARCHER—the same fellow who watched a collapsing turbine destroy his friend's Nikon a few decades ago—"and it slung some aluminum sheet metal into the university president's front yard." In better news, its turbines all survived sustained winds of seventy miles per hour in 2011. NOLAN CLARK has retired and is working on a book about small wind machines, and the Department of Agriculture's wind research program in Bushland has ended, though some of its work will continue at nearby Texas Tech University.

MICHAEL OSBORNE works on renewable energy for Austin Energy, dreaming up the next great thing, whether it's wind (which he thinks will max out at around 20 or 30 percent of electricity use), solar, or anything else. "What we're doing right now will look just like, you know, hay in the barn for the horses," he says. "It will be kind of quaint."

JOE JAMES still hears confessions at the Mercy Center in Slaton, near Lubbock. He helped renovate the center with his own hands and now takes free lodging there. "I don't feel like I'm bumming on anyone," he says.

BOB KING heads Good Company Associates, and DALE OSBORN is president of a Colorado company called Disgen that specializes in small-scale, or "distributed," power projects. "For rural Texas and all rural communities in America, renewables can be the economic growth engine desperately needed," Osborn says. MARK ROSE, formerly of the LCRA, promotes smart-grid initiatives from the helm of a Bastrop-based electric co-op called Bluebonnet, while TOM FOREMAN left the LCRA in 2012 for another post.

WALT HORNADAY continues to develop ever-larger wind farms, including several in the Panhandle; one Texas project he worked on was so ambitious that in 2009 it caught the eye of New York senator Chuck Schumer, who denounced the planned use of federal stimulus money to buy Chinese turbines. ANDY BOWMAN and PATRICK WOODSON work on renewables in Texas, as does WARD MARSHALL.

SAM WYLY made a mint with Green Mountain Energy before it was purchased by a big power company called NRG Energy for $350 million in September 2010. That was a few months after Wyly and his brother, Charles, were accused by the Securities and Exchange Commission of reaping more than $550 million from insider trading and fraud (Sam Wyly is fighting the charges; Charles is now dead). By that time Sam Wyly had more than given back to the governor who had helped him out with the 1999 deregulation measure by backing ads that trashed his rivals for the presidency in 2000 and 2004, John McCain and John Kerry, respectively. "At the time we genuinely believed there'd be a lot stronger [environmental] action if Bush was going to be in Washington," Wyly explained later. "He had other priorities once he was in office."

GEORGE W. BUSH, for his part, was widely vilified by the environmental movement during his presidency, not least for his failure to push for a renewable-energy mandate similar to the one that energized Texas wind in 1999. Still, the wind industry gives him plaudits for past deeds. When Bush keynoted the American Wind Energy Association conference in Dallas in 2010, he drew a standing ovation. "There's a big difference between the talkers and the doers," Bush told them, "and here in the state of Texas, we are doers."

Despite the 1999 deregulation triumph, DAVID SIBLEY narrowly lost a senate vote to become lieutenant governor when Rick Perry assumed the governorship following Bush's election to the White House. Sibley retired from the senate in 2002 to set up a lobbying shop. He dropped back into the public eye in 2010 to run for his old seat, but in a year in which voters were not all that interested in hefty legislative accomplishments, lost the Republican primary to a Tea Party candidate. PAT WOOD accompanied Bush to Washington, where he chaired the Federal Energy Regulatory Commission for four years before heading back to Houston to develop transmission and other energy projects. He has considered whether solar should have special incentives, like wind, but "when I look at the whipsaw around the world, as policies have been yanked out from underneath the industry, I'm thinking we'd probably just be better moving forward on the normal economics," he says.

TOM "SMITTY" SMITH and JIM MARSTON still can be found in the halls of the Texas capitol, searching for allies for their environmental causes. STEVE WOLENS has returned to private law practice in Dallas after leaving a twenty-four-year-career in the Texas house. His work has nothing to do with wind, but his wife, Laura Miller, served as Dallas mayor from 2002 through 2007, where she concentrated on cleaning up North Texas's skies. "It took my wife's efforts as mayor to do a mop-up of what I didn't get done in the legislature," Wolens says. In 2003 Wolens hopped on a plane with former Enron executives to fly to West Texas. "I was just shocked at how many windmills we had out there," he says. "Little did I know how quickly we would get to it and exceed it."

In August 2012 JAY CARTER SR., who had teamed with his son to build some of Texas's first wind turbines, passed away. As for the old Carter warehouse in Burkburnett, it looks like a sepulcher. One spring afternoon in 2011, a couple of flaky wind blades lay out back behind the building. Inside, a fifty-foot-long blade mold was covered in dust; supported by a network of crosshatched legs, it could be mistaken in the half-darkness for a dinosaur spine. JAY CARTER JR. still roams the place, but a new company, Carter Wind Energy, is now led by Jay Jr.'s fresh-faced son, MATT CARTER, who swept the blade shop floors in junior high and spent college summers on turbine maintenance crews. Matt hopes to revive the wind business and hand it over to his children, though the next generation of turbines will likely be built out of state, perhaps overseas, where labor is cheaper. The Carter company now specializes in research and the promotion of its design. "We just believed in the idea of wind," says Matt, a Texas Tech engineering graduate like his father. The family business, he adds, "came down to perseverance."

NOTES

INTRODUCTION

There is a wide literature about Texas and wind. In this chapter and the book generally, we've relied on beautiful memoirs by A. C. Green (*A Personal Country*) and Gail Caldwell (*A Strong West Wind*), as well as Lou Halsell Rodenberger et al.'s *Writing on the Wind*, a compilation of female West Texas authors that includes an essay by Lisa Sandlin, cited in this chapter.

Fiction, too, helps to evoke the places. The quiet novels of Larry McMurtry helped us think about the Panhandle and West Texas. On the other end of the spectrum lies the wonderfully, hysterically melodramatic novel *The Wind*, by Dorothy Scarborough.

Throughout the book, we've looked to statistics kept by a variety of sources, including the American Wind Energy Association (AWEA), the Railroad Commission of Texas, the Energy Information Administration, the U.S. Census, and ERCOT, the state grid operator.

Articles quoted in this chapter include "Something New under the Sun," *Dallas Morning News* (June 15, 1986): 30A (no author given); Jennifer Bogo, "The New Wildcatters," *Popular Mechanics* (December 2009): 92, published on PopularMechanics.com, pp. 91–130 (http://www.heliovolt.com/files/the-new-wildcatters.pdf; Michael Webber's comments); and "Wind Power in Texas: Blowing Strong," *The Economist*

(June 29, 2006) (Jerry Patterson's quotation). In this chapter and throughout this book, we also draw from Kate Galbraith and Asher Price, "A Mighty Wind," *Texas Monthly* (August 2011).

Information on Texas geology is found in *Texas: A Guide to the Lone Star State* (New York: Hastings House, 1940), and correspondence with geologist Chock Woodruff. Georgia O'Keeffe's quotation, "terrible winds and a wonderful emptiness," comes via Laura Bush's memoir, *Spoken from the Heart* (New York: Scribner, 2010). Robert E. Lee's comments on the wind come from a letter to his daughter, quoted in Elizabeth Brown Pryor, *Reading the Man* (New York: Penguin Group, 2007), p. 249. The Coronado quotation is found in the "Llano Estacado" entry in the *Handbook of Texas Online*, a resource that was immensely helpful throughout this book (http://www.tshaonline .org/handbook/online). The Richards quotation ("I thought I knew" is cited in Amy Schatz, "Remembering Ann Richards," *Wall Street Journal* (Washington Wire) (September 14, 2006).

Interviews used in this chapter include those with Chris Crow, Randy Sowell, and Patrick Woodson.

1. FOLLOWING A GLIDER

Information on Michael Osborne's life comes via several interviews and e-mails with Osborne from 2010 to 2012. His book, *Beyond Light and Dark*, provides supplementary recounting of his glider experience (pp. 110–111), as well as his broader philosophy. *Gray County Heritage*, by the Gray County History Book Committee (Dallas: Taylor Publishing Co., 1985), provided supplemental Osborne family history.

We relied on a variety of sources for descriptions and history of the Texas Panhandle. These include Mark J. Stegmaier, *Texas, New Mexico, and the Compromise of 1850*; S. C. Gwynne, *Empire of the Summer Moon*; Caldwell, *A Strong West Wind*; Frederick Rathjen, *The Texas Panhandle Frontier*; Roxana Robinson, *Georgia O'Keeffe* (source of "Sometimes, when" quotation, p. 89); Timothy Egan, *The Worst Hard Time*; and Nellie Witt Spikes, *As a Farm Woman Thinks*.

Growth and elevation figures for early Pampa are drawn from the *Handbook of Texas Online* ("Pampa"), and the handbook's "Dugouts" entry helped with descriptions of early settlements. Bill Neiman offered descriptions of native Panhandle vegetation. Woody Guthrie's comments about the Red Sea and the Dust Bowl come from a 1940 oral history conducted by Alan Lomax for the Library of Congress Archives, at http:// soundportraits.org/on-air/woody_guthrie/transcript.php, and we also drew from Ed Cray, *Ramblin' Man*, and Guthrie's autobiography, *Bound for Glory* ("wilder than a woodchuck," pp. 162ff). Guthrie's complete lyrics are found at www.woodyguthrie.org.

The description of pioneers' wistful attempts at wind wagons comes from a 1978 publication by Ray Pierce, *Wind Energy*, found in Coy Harris's library.

For general descriptions of water-pumping windmills we drew extensively from Robert Righter's comprehensive *Wind Energy in America*. Also helpful were Richard Leslie Hills, *Power from Wind*; Peter Asmus, *Reaping the Wind* (pp. 25–28, including the 100,000 windmills figure); and Baker, *American Windmills* (especially pp. 26 and 53) and *A Field Guide to American Windmills*.

We learned about the use of windmills by the Dutch resistance from the obituary "Hilda Van Stockum, 98, Prolific Children's Author," Stephen Miller, *New York Sun* (November 3, 2006), and from *Painting in the Dutch Golden Age*, published in 2007 by the National Gallery of Art, Washington, D.C. (p. 10), which also describes millers' use of windmills as signals of personal change, p. 11.

A piece by Baker in *Windmillers Gazette* (Winter 1983): 2 supplied the detail on Halladay's work at Harper's Ferry, as well as the quotation, "Now one of the greatest," which Baker got in turn from *The Past and Present of Kane County, Illinois* (Chicago: Wm. LeBaron, Jr. & Co., 1878, p. 598).

On Texas windmills in particular, J. Evetts Haley, *The XIT Ranch of Texas and the Early Days of the Llano Estacado*, was invaluable, as was a 1963 journal article by Terry Jordan titled "Windmills in Texas" (p. 81). The *Handbook of Texas Online* describes the sixty-mile railway from Houston, and the entry on windmills cites Texas as the largest source of demand for windmills.

The *Handbook of Texas Online* (Midland) was the source for the quotation, "virtually every house." The W. P. Gillespie quotations are from *Windmillers Gazette* (Summer 1982): 9; the *Gazette* in turn reprinted the article from *Descriptive Catalogue of U.S. Wind Engine & Pump Co.*, produced by U.S. Wind Engine and Pump Company, Batavia, Illinois (Chicago: Rand, McNally & Co., c. 1876), p. 13.

The windmills-pumping-oil piece is from Baker's *American Windmills* (p. 80). "Within a short time" is from Walter Prescott Webb's *The Great Plains* (p. 336); "It was the acre or two" is from p. 346. A fall 1900 Sears Roebuck & Co. catalog (Northfield, Ill.: DBI Books), p. 944 (from Coy Harris's collection) helped us understand windmill pricing.

On the XIT Ranch, we drew from Haley, *The XIT Ranch* (pp. 95, 167); the Web site http://www.xitmuseum.com/history.shtml ("largest range in the world under fence"); "J. B. Buchanan's Windmill Collection," by Paul Gipe, *Wind Power Digest*, (Fall 1979): 48; and the *Handbook of Texas*'s XIT Ranch entry. Jon Naar, *The New Wind Power* (p. 64), supplied further information and quotations on the windmill experiences of J. B. Buchanan. The Tater Crouch episode comes from Proulx, *That Old Ace in the Hole*, p. 146. "Best market area for windmills" is from Jordan, "Windmills in Texas" (p. 83); the farmers' wives tale is from Baker, *American Windmills* (p. 68). Descriptions of Thomas O. Perry's work come from Hills, *Power from Wind*; and Gipe's *Wind Power*. The timeline on Aermotor's Web site (http://www.aermotorwindmill.com/Company /History.asp), as well as Jordan, "Windmills in Texas," helped with the history of the Aermotor.

Web sites from the National Oceanic and Atmospheric Administration (NOAA) and the National Weather Service provided elevation and rainfall information for Pampa and the High Plains.

On the music scene in Lubbock, we drew on the Buddy Holly Center's Web site (http://www.buddyhollycenter.org/gallery/biography.aspx) for basic biographical information (including "play for the opening"), and we got a quotation from Butch Hancock from Virginia Raymond's master's thesis, "'The Wind' in the Literary Creation of West Texas" (p. 136). The Michael Osborne quotation, "we had a few songs," comes from *Beyond Light and Dark* (p. 110). Rick Koster's *Texas Music* set the scene for Austin in the 1960s and 1970s, and a Web site dedicated to the Armadillo (http://www.armadilloworldheadquarters.com/awhq.htm) provided context, including the manifesto.

On Osborne's 1981 wind farm in Pampa, sources include interviews with Osborne; the J. C. Andrews short film, *Lease the Wind*, lent by Michael Osborne, which documents the creation of that wind farm; an August 2011 *Texas Monthly* article, "A Mighty Wind," by Price and Galbraith (some descriptions are taken directly from that piece); and (for Osborne's solar background) "Something New under the Sun," and "Renewed Effort for Renewable Energy," by David Butts, *The Texas Observer* (August 31, 1984): 10–11. A quotation from Woody Guthrie ("roar around in the wind") comes from his autobiography, *Bound for Glory*. Asmus's *Reaping the Wind* was our main source on the Crotched Mountain wind farm in New Hampshire.

Descriptions of Pampa and T. Boone Pickens's plans come from "A Mighty Wind," by Price, *Austin American-Statesman*, November 2, 2008.

2. THE TINKERERS

A rich literature exists on windmills and windmilling. We drew once again from the *Handbook of Texas Online* ("windmills"); Haley, *The XIT Ranch*; Anne Dingus, "The Windmill"; Spikes, *As a Farm Woman Thinks*; and interviews with Nolan Clark and Coy Harris. Texas census figures come from http://www.census.gov/population /censusdata/urpop0090.txt, and quotations from Proulx, *That Old Ace in the Hole*, are on pp. 141, 126, 154, and 146. The Webb quotation comes from *The Great Plains* (p. 320). Ken Starcher of AEI helped with the explanation of wind speeds and mesas.

For context on the XIT Ranch's set of windmillers, we used, once again, Haley, *The XIT Ranch*, and Baker's books. The J. B. Buchanan comments come from "Lone Star Wind," *Windmillers Gazette* (Fall 1979): 48. T. Lindsay Baker's contributions come from "Unusually Tall Windmill Towers," *Windmillers Gazette* (Summer 1989): 2.

Interviews in 2011 with Bob Bracher of Aermotor in San Angelo and Mike Crowell in Claude supplied the history of their respective operations.

In describing the nature of the wind in Texas, *Texas Weather*, by George W. Bomar, was helpful (especially p. 181), as was Donald R. Haragan's *Blue Northers to Sea Breezes* and the *Handbook of Texas*'s entry on Blue Northers. We drew, clearly, from Cormac McCarthy's *All the Pretty Horses*, and from the poem "Wind," from Larry D. Thomas's collection *Where the Skulls Speak Wind*.

On the technology of modern wind-electric turbines, our sources included "Wind Power Gains as Gear Improves," by Kate Galbraith, *New York Times* (August 7, 2011). Hills's *Power from Wind* was also helpful, especially p. 271 on air density.

For accounts of Moses Farmer and Charles Brush, we turned to Righter, *Wind Energy in America* (esp. pp. 38, 42–58). Lord Kelvin's work is easily found online in an 1881 issue of *Science*, vol. 2, by British Association for the Advancement of Science (p. 475). Naar's commentary is from *The New Wind Power* (pp. 47–48). LBJ's "Flying Windmill" is chronicled in Robert Caro's *The Years of Lyndon Johnson*, vol. 2.

Descriptions of Marcellus Jacobs come from several sources, including Righter, *Wind Energy in America*; a 1973 interview with Righter in *Plowboy*, reprinted by *Mother Earth News* online at http://www.motherearthnews.com/Renewable-Energy /1973-11-01/The-Plowboy-Interview.aspx; a 1982 article by Shoshana Hoose in the *Minneapolis Star* titled "Wind Power Propels Family's Trade" (the specific date is not

given, but it was accessed at http://www.jacobswind.net/wp-content/uploads/2011/02
/Minneapolis-Star-1982.pdf); Naar, *The New Wind Power* (pp. 47–48, also helpful for
wind-electric background); and Claude Eggleton, "Marcellus L. Jacobs (1903–1985)."

Adam Holman, while working in Bushland for West Texas A&M University, kindly
described in a July 29, 2011, e-mail his grandfather's experiments with a self-carved
wind machine. Hal Phelps, in interviews in 2011 and 2012, described his boyhood curi-
osity about wind chargers. The quotation from Robert Rodale on the wind machine
numbers comes from Naar, *The New Wind Power* (p. 15); Spikes's memoir, *As a Farm
Woman Thinks* (p. 232), was also helpful.

Statistics on the percentage of Texas farms with pre–Rural Electrification Administration wiring come from "REA Begins Bringing Electricity to Rural Texas," an online
article from the Texas State Historical Association (http://www.tshaonline.org/day-by
-day/31054). Thomas Edison's flirtations with windmills are documented in Edwin
Black, *Internal Combustion* (p. 128).

The best description of what rural electrification meant in Texas, particularly in
the Hill Country, comes from *The Path to Power*, part of Robert Caro's LBJ series (see
p. 516 for the $5,000 cost per mile and the 6 of 6.8 million farms figures). For infor-
mation on Bartlett, the first Texas community touched by the REA, we used two online
resources: http://www.bartlettec.coop/History.aspx and http://www.texascooppower
.com/energy/electricity-basics/electricity-basics/historic-connection"; the latter is an
October 2010 article titled "Historic Connection in Texas Co-op Power," by Charles
Boisseau. The *Handbook of Texas*'s "REA Begins Bringing Electricity to Rural Texas"
was helpful on figures and on the REA arriving in Deaf Smith County.

On J. B. Buchanan, an article by the man himself, "Correspondence: J. B. Buchanan
and His Lifelong Love of Windmills," in *Windmillers Gazette* (Spring 1983) was helpful, as was Gipe, "Lone Star Wind." Other details, such as how farmers were forced to
take down their wind machines before the REA would deign to reach their property,
are from Gipe, "Stockett's Wind Wheel," *Wind Power Digest* (Fall 1979): 46. Spikes, *As
a Farm Woman Thinks*, notes that the wind could blow out fuse boxes (p. 110).

For the wonderful tale of Joseph Spinhirne, we drew from Naar, *The New Wind
Power* (pp. 70–72) as well as Ken Starcher's recollections. The Righter quotation, "no
wind-electric company survived," comes from *Wind Energy in America* (p. 105). The
notion of a wind tower as a perch for a TV antenna comes from the same book (p. 125).
The 1965 figures for wind farms without electricity come from the *Handbook of Texas
Online* ("REA Begins Bringing Electricity to Rural Texas"). The sorry fate of the Brush
windmill comes from Righter, *Wind Energy in America* (p. 57).

Finally, information about Joe James and his father, Andy Marmaduke James,
comes from a series of interviews in person and over the telephone with Joe James in
2010–2012.

3. THE OIL EMBARGO

Notes from Milton Holloway (the Austin newcomer quoted), David Blackmon (Beeville; via Twitter, September 19, 2011), and NOAA (Lubbock and Austin) supplied
weather data and stories for January 1973.

The executive order creating the Governor's Energy Advisory Council can be found at http://repository.law.ttu.edu/bitstream/handle/10601/1022/DB2(1973).pdf ?sequence=1; it contains related Texas energy statistics, as does *Texas Renewable Energy Resource Assessment*, prepared by Virtus Energy Research Associates for the TX SEDC (p. 18 for the quarter of crude oil data point). Interviews with Bill Hobby Jr., Max Sherman, and Milton Holloway were helpful. Railroad Commission statistics, found easily on the agency's Web site, provide well-count and production information on Texas oil and gas. The factoid that gas heated about half of U.S. homes during the 1970s comes from Freeman, *Energy* (p. 21). The 90 percent of Texas electricity statistic comes from "Texas Power Companies Converting from Natural Gas to Coal, Lignite," *Science* (November 4, 1977).

For the Coastal States debacle and the energy shortages in Texas, we drew extensively from Paul Burka's lengthy *Texas Monthly* article, "Power Politics." David J. Lynch, "Prison Term Could Cap Oil Trader's Legendary Career," *USA Today* (December 22, 2005), filled out Oscar Wyatt's biography.

Information on Max Sherman comes from a 2011 in-person interview.

As to the tumultuous aftermath of October 1973, we drew statistics and commentary from Freeman's *Energy*, which was written in 1974 as a quick reaction to the oil embargo ("energy Pearl Harbor Day" is a quotation from p. 3; also helpful were pp. 4–5 and 37) and Olien and Olien, *Wildcatters* (p. 142; university lands statistics are at p. 144). Railroad commissioner Mack Wallace's comments come from "Mack Wallace Speaks Out About the Oil Import Fee," *Quorum Report* 4, no. 11 (February 1, 1986): 2. The Nixon quotation is widely available.

Texas documents on the energy-crisis era, including newspaper articles and farmers' pamphlets, come from the Dolph Briscoe papers, housed at the Dolph Briscoe Center for American History at the University of Texas at Austin—including the clip referencing "exotic fuel sources." Some commentary has already appeared in "Texas' Energy Lessons from the 1970s," by Kate Galbraith, April 4, 2011. The term "solar wind farm" comes from Butts, "Renewed Effort for Renewable Energy," *Texas Observer* (August 31, 1984): 10–11.

In writing about the Smith-Putnam wind machine in Vermont, we drew from Asmus, *Reaping the Wind* (pp. 44–46); the Lewis Research Center pamphlet, *Wind Energy Developments in the 20th Century* (pp. 2–3); and a 1973 talk in Washington, D.C., by Beauchamp E. Smith, retired president of the S. Morgan Smith Company, available online as part of the "Wind Energy Conversion Systems: Workshop Proceedings" (pp. 6–7). Hills, *Power from Wind*, contributed the "meager and uncertain" quotation (p. 274). Some of these sources helped out with the description of Percy Thomas's wind plans (a talk specifically on Thomas was given at the 1973 Washington wind conference and is available online), but information on Thomas is also nicely summarized in a February 1947 *Popular Mechanics* article, "Can Cities Harness the Wind?" by John L. Kent (available online).

Statistics on federal wind funding (the $300,000 and $67 million figures) come from Vaughn Nelson, *Wind Energy* (p. 217); and a summary of the Lockheed study comes via Robert King, *Alternatives to the Energy Crisis* (p. 153). Lewis Research Center, *Wind Energy Developments*, was a crucial source on big industrial companies' entry into wind.

A telephone interview and follow-up e-mail with Bill Hobby Jr. elicited the story on

Gov. Dolph Briscoe's boyhood wind machine, and interviews with Earl Gilmore and Vaughn Nelson fleshed out how those two banded together.

The stats on Amarillo wind speeds come from the Nelson and Gilmore presentation at the June 1973 wind workshop in Washington, D.C. (information online, p. 33), but most other information comes from the 1974 Nelson and Gilmore report itself, *Potential for Wind Generated Power in Texas* (kindly lent to the authors by Milton Holloway).

Our section on the Jay Carter family comes mainly from a visit and interviews, with follow-up e-mails and calls, with the Carters themselves (primarily Jay Carter Jr.). Additional information, including quotations from Jay Carter Sr., is from the 2011 *Texas Monthly* short video, *Three Generations of Texas Wind Men*, http://www.texasmonthly .com/multimedia/video/home/15902). We also drew from the October 1974 *Popular Science* article on the Carters.

Information on work at Bushland comes from interviews with Nolan Clark and *Handbook of Texas Online* entries on Bushland, barbed wire, and the Frying Pan Ranch; and an online interview with Nolan Clark from July 1, 2003, found at a Department of Energy Web site, Wind Powering America, http://www.windpoweringamerica.gov /filter_detail.asp?itemid=685, was also helpful. Interviews with Kenneth Starcher and Vaughn Nelson described developments at Bushland and also West Texas State University. The pamphlet *Getting the Most from Every Drop: A Checklist for Saving Farm Fuels* was found in the Dolph Briscoe collection at the University of Texas at Austin.

Information on Father Joe James's work comes from an in-person interview, supplemented by reading in *Spectra*, the one-time magazine of the Texas Solar Energy Society, whose 1982–1983 issues helped nail down dates for the James project. Articles in the *Lubbock Avalanche-Journal* detail the Lubbock miracles and James's demotion, including "Faith That Endures," by Beth Pratt, August 9, 2003 (http://lubbockonline .com/stories/080903/rel_0809030074.shtml), and "Visitors Flock to Assumption Masses for Healing, Revelation," by Beth Pratt, August 16, 2003 (http://lubbockonline .com/stories/081603/rel_081603082.shtml). See also "Visions of a 'Miracle,'" by Doug Hensley, for the *Avalanche-Journal*, in the 2009 online commemoration of the Lubbock centennial, http://www.lubbockcentennial.com/AJremembers/081008.shtml, and "Monsignor Joseph James Celebrates 50 years," at www.mercymessenger.com /JWJ_bio.pdf. Interviews with Coy Harris were also helpful, and an interview with Randy Sowell yielded the wonderful anecdote about boys biking with sails.

4. THE 1980S: BOOM—THEN BUST

To recap Osborne's wind farm, we drew on interviews with the players, as well as Andrews's film *Lease the Wind*.

Some of the most comprehensive facts and figures on the 1980s' wind roller-coaster are from an unpublished source, Janet Sawin's lengthy Fletcher School dissertation, "The Role of Government in the Development and Diffusion of Renewable Energy Technologies," which she e-mailed to the authors. Facts we drew from it include wind development peaking at $186 million (p. 112).

The quotation "Make no mistake about it" comes from "Windfarming in America," by Mike Evans, *Wind Power Digest*, no. 24 (Summer 1982): 6–12.

The "known universe" quotation about Osborne's wind farm is from "Something

New under the Sun," *Dallas Morning News*. Osborne supplied the 2.69 cents per kwh figure. The Dallas newspaper quotation ("Kind of fun") is from "Entrepreneurs Making Money While the Sun Shines," by Tom Lindley, *Dallas Times Herald* (June 8, 1982): 1, 4.

We owe thanks to our *Texas Monthly* fact checker, Valerie Wright, who helped us obtain additional details on Joe James (such as the $35,000 repayment over three years data point).

A variety of government Web sites and books, as well as Sawin, "The Role of Government," lay out the push and pull of various incentives in the 1980s.

For Crotched Mountain's history, once again we drew from Asmus, *Reaping the Wind* (pp. 61–62). As for hazards of the oil fields, we turned to Olien and Olien, *Oil in Texas* (p. 68). The anecdotes about wind machines on Gulf oil rigs come from "Can We Harness the Wind?" by Roger Hamilton, *National Geographic* 148, no. 6 (December 1975): 812–828. Details of Marcellus Jacobs's getting back into the game come from "Wind Power Propels Family's Trade" (pp. 3, 10), at http://www.jacobswind.net/wp-content/uploads/2011/02/Minneapolis-Star-1982.pdf. Another source on Jacobs, including the quotation about old Jacobs machines being dug out, cleaned, and put back into use, is "Marcellus L. Jacobs (1903–1985)," by Eggleton.

Tales of Hummingbird Wind Power come mostly courtesy of Carlos Gottfried, interviewed in November 2011, as well as from a 2011 e-mail from Caroline Crimm, a former Hummingbird employee.

Tales of the Alternative Energy Institute and its early mishaps come mostly from Ken Starcher at AEI, as well as Vaughn Nelson. The Carter quotation, "We've had some units," comes from Andrews, *Lease the Wind*. The comment from José Zayas of Sandia came during a 2011 interview with Kate Galbraith.

Information on crude oil prices in the '80s comes from the Department of Energy. A variety of sources, including many news clips from the Dolph Briscoe collection; Sawin, "The Role of Government"; and David Yergin's energy classic, *The Prize*, helped flesh out the policy swings of the 1980s. The Railroad Commission of Texas provided information on its 1979 repeal of the natural gas boiler rule. The tale of Billy Jack Mason's free gas comes from "Even Texans Can't Resist Gas at 0 Cents a Gallon," by David Maraniss, *Washington Post* (April 4, 1986): A3.

Reagan's comment, "The Department of Energy has a multibillion-dollar budget," comes from "Where Did the Carter White House's Solar Panels Go?" by David Biello, August 6, 2010, at http://www.scientificamerican.com/article.cfm?id=carter-white-house-solar-panel-array. The Harvard Business School did a thorough and helpful review of wind: "Historical Trajectories and Corporate Competencies in Wind Energy," by Geoffrey Jones and Loubna Bouamane, May 2011, at http://www.hbs.edu/research/pdf/11-112.pdf. Evans, "Windfarming in America," presents the struggles of big companies like Boeing. Alcoa's misfortunes are chronicled in Gipe's *Wind Energy Comes of Age* (p. 361) and Righter's *Wind Energy in America* (p. 172).

Information on the American Wind Energy Association conference in Amarillo in 1982 comes from "The Wind Industry Has Matured," by Joseph Deahl, *Wind Power Digest*, no. 25 (Fall 1982). The 1988 U.S. R&D figure is from Nelson, *Wind Energy* (p. 220), and Sawin, "The Role of Government," p. 375, helped with 1982.

Some basics on California's wind resources and ranking come from an American Wind Energy Association fact sheet, http://www.awea.org/learnabout/publications/upload/1Q-11-California.pdf, and also the May 9, 2005, paper "California Wind Resources," by Dora Yen-Nakafuji, at http://www.energy.ca.gov/2005publications/CEC -500-2005-071/CEC-500-2005-071-D.PDF.

Brown's "Woodchips and Windmills" moniker is cited in "As Young Governor, Brown Went His Own Way," by Evan Halper, *Los Angeles Times* (October 29, 2010). Stats on oil and gas as 70 percent of California's energy production come from Sawin, "The Role of Government" (p. 186). Sawin (p. 224) also gives California's total wind capacity.

Recollections of the Carters in California come from Jay Carter Jr. and Asmus, *Reaping the Wind* (p. 13). Megawatt data on California's wind buildout, by year, is from Sawin, "The Role of Government," p. 607. On p. 238, Sawin explains how wealthy investors could get large amounts of their money back, and on p. 205 she discusses the $100 million the state doled out in tax credits (citing a California Energy Commission report).

Bob King's background comes from interviews and e-mails with him; the quotation from Ed Vetter comes from *Bill Clements*, a biography by Carolyn Barta (p. 236).

The Carters' fortunes are told mostly by Jay Jr.; the skeptical note regarding the Montana contest sounded by Righter comes from *Wind Energy in America* (p. 254).

Matthew Wald, "A New Era for Windmill Power," *New York Times* (September 8, 1992), describes California wind as having "generated more tax credits than electricity." Asmus's analysis of Tehachapi comes from *Reaping the Wind* (pp. 101–103). Nelson's comment, "Wind machines that never were," is found in J. Deahl, "The Wind Industry Has Matured" (pp. 48, 52). The Russel Smith comment, "A lot of the wind schemes," can be found in Butts, "Renewed Effort for Renewable Energy." "Can We Harness the Wind?," *National Geographic*, expresses fears of slowing winds. California bird death statistics, studies, and commentary come from Paul Gipe's *Wind Energy Comes of Age* (pp. 344–345).

Final Osborne commentary, on his solar ambitions, comes from interviews with him as well as "Something New under the Sun," *Dallas Morning News*. An interview with Milton Holloway helped with the discussion of the Texas Energy Development Fund and the fate of the Governor's Energy Advisory Council.

5. ANN RICHARDS—AND
A BIG WIND FARM AT LAST

For material on energy in the Ann Richards administration, we drew on archives at the Dolph Briscoe Center for American History, especially the papers of John Fainter, her former chief of staff. An interview with Carol Tombari gave us a sense of the inner workings of the Richards administration on alternative-energy issues. Articles helpful on the Richards front (and Texas energy issues during the early 1990s) include "The Daughter Also Rises," by S. C. Gwynne, with reporting by Michael Hardy, *Texas Monthly* (August 2004): 112; "This Texan Is a Straight Shooter," David Nyhan, *Boston Globe* (October 1, 1989): A29; "Richards: Gulf War 'about Oil'; U.S. Needs Policy on Energy, She Says," Clay Robison, *Houston Chronicle* (June 17, 1992): A19; "Awash in Imports, the Oil Business Dwindles," by Matthew L. Wald, *New York Times* (July 26, 1992): 6.

We consulted the State of Texas Energy Policy Partnership, vol. 1, Report to the Governor, the Legislature, and the Citizens of Texas, March 1993 (http://www.seco .cpa.state.tx.us/zzz_seconews-links. . ./stepp-vol1-1993.pdf). An interview with Karl

Rábago helped on this front, too, as did *Texas Renewable Energy Resource Assessment*, prepared by Virtus Energy Research Associates, Mike Sloan's shop. The newsletters from the Texas Renewable Energy Industries Association, impeccably cataloged by Russel Smith, provided important contemporary commentary. The Union of Concerned Scientists study and the Stavins quotation come from "New Studies Predict Profits in Heading Off Warming," by William K. Stevens, *New York Times* (March 17, 1992): C1. The source for the oil and gas contribution to gross state product is the Texas comptroller's Web site, http://www.window.state.tx.us/specialrpt/energy/nonrenewable/ex hibits/exhibit4-2.php. Information on the 25MW Minnesota wind farm opening in 1994 comes from "Buffalo Ridge Officially Dedicated," *Windpower Monthly* (August 1, 1994).

Kenetech information draws heavily from Asmus's reporting in *Reaping the Wind*, which includes quotations from Dale Osborn ("This is not a fraternity" and "can't afford to look like Governor Moonbeam's children"). We also interviewed Dale Osborn and consulted a variety of other sources: "Kenetech Rides the Foreign Trade Winds," by Steve Ginsberg, *San Francisco Business-Times* (January 27, 1995): A7; "'Farm' Harvests the Wind for Electric Power," by Maria Lenhart, *Christian Science Monitor* (January 27, 1982): 17; and "Wind Company Blows in with $72 Million IPO," by Clifford Carlsen, *San Francisco Business-Times* (July 23, 1993): 3. Information about Kenetech's downfall comes from "Taking the Wind Out of Sales; Hurricane-force Gusts Damage Turbines at Electricity Plant," by Barry Shlachter, *Fort Worth Star-Telegram* (January 25, 1996): 17, and "Gone with the Wind? Kenetech Insiders Brace for Worst," by Herb Greenberg, *San Francisco Chronicle* (December 12, 1995): C1. "Marketplace; A Giant of Wind Power Stumbles Badly," by Agis Salpukas, *New York Times* (December 27, 1995): D1, also helps explain Kenetech's downfall and the suits it faced, as does "Kenetech's New Turbine Has Start-up Problems but Utilities Not Concerned," *Independent Power Report* (December 16, 1994): 9, and "Blowing in the Wind," by Kambiz Foroohar, *Forbes* (December 4, 1995): 14.

Information on Ed Wendler comes from "Ed Wendler's Rise, Fall and Rise Again," Kyle Pope, *Austin Business Journal* 7, no. 38 (October 26, 1987), sec. 1: 9, and "Ed Wendler Sr. 1932-2004: Austin Lobbyist Was Respected Political Guru," Dick Stanley, *Austin American-Statesman* (March 7, 2004): A1.

Interviews and or e-mails with Robert Cullick, Tom Foreman, Bob King, Tony Kunitz, Garry Mauro, Dale Osborn, Michael Osborne, and Mark Rose explain the relationship between the LCRA and Kenetech and the history of the Delaware Mountains wind farm. The Texas Renewable Energy Industries Association newsletters from 1995 and 1996 were invaluable here. We also quote from "Kenetech Signs Pact for Texas Wind Plant," UPI, November 29, 1993, and "Power Blowin' in the Wind; Site in Remote West Texas Sprouts Turbines That Help Provide Austin Homes with Electricity," by Bruce Hight, *Austin American-Statesman* (October 1, 1995): D1, which includes the $40 million figure; and "Turbines Turn West Texas Wind into Power," by John MacCormack, *San Antonio Express-News* (September 17, 1995) (this includes many details such as the height of the site). We also used figures from Sawin, "The Role of Government," and we quote from *The Path to Power*, the first volume of Robert Caro's LBJ biography.

LCRA background comes from its Web site, as well as from Jimmy Banks and John E. Babcock's 1998 history, *Coralling the Colorado*. Quotations from S. David Freeman appear in "LCRA Leans on Experience with New Leader," by Asher Price, *Austin American-Statesman* (June 30, 2011), which cites a 1986 *Statesman* quotation, and from a July 23, 2007, interview with Freeman for the Texas Legacy Project (available online at http://www.texaslegacy.org/m/transcripts/freemandavidtxt.html). The fall 1995 TREIA newsletter contains "An Open Letter to TREIA from Texas Land Commissioner Garry Mauro" (p. 6), plus the $3 million projections for the Permanent School Fund; the summer 1995 issue ("Texas Wind Power Plant Generates Electricity for the First Time," by LCRA News), contains information on the starting date.

Additional material on the Delaware Mountains windstorm damage and aftermath comes from the TREIA newsletters. The "like matches" comment comes from "Texas Windpower Project Nears Full Operation after Surviving One-in-a-Million Wind Storm," *TREIA Newsletter* (Winter 1995): 10. Also useful were "Heavy Winds, Twisters Cause Death, Damage" (January 18, 1996), by Associated Press, in *Austin American-Statesman* (p. B5); "Blowin' in the Wind Ain't Such Fun on Highway 61," Phil Davison, *The Independent* (January 24, 1996): 11; "Tornadoes, Winds Rake N. Texas," Brian D. Crecente and Bill Hanna, *Fort Worth Star-Telegram* (January 18, 1996): 1; "An Ill Wind Blew No One Any Good," by Bruce Hight, *Austin American-Statesman* (January 27, 1996). The chapter closes with information from "Wind-Power Brings Schools $29,365," *Austin American-Statesman* (March 2, 1996): B2.

6. WINDCATTERS

Articles cited in this chapter include "Big Spring Embraces Big Wind," *TREIA Newsletter* (Fall 1998); "'Farm' Harvests the Wind for Electric Power," by Maria Lenhart, *Christian Science Monitor* (January 27, 1982): 17 ("new kind of cash crop"); "FPL Energy Dedicates Texas' Largest Wind Plant," *TREIA Newsletter* (Spring/Summer 1999): 4.

"Roping the Texas Breezes," SECO (State Energy Conservation) Fact Sheet, no. 14 (http://www.infinitepower.org/pdf/FactSheet-14.pdf) was helpful for facts and figures about the late 1990s wind projects. Sawin, "The Role of Government," supplied the facts on Denmark's getting 15 percent of its electricity from wind by 2001 (p. 22).

Interviews with Andy Bowman, Robert Bradley, Mark Bruce, Jay Carter Jr., Chris Crow, Sam Enfield, Walter Hornaday, Ward Marshall, Michael Osborne, Mike Sloan, Delbert Trew, and Patrick Woodson were helpful for this chapter.

Enron information comes from "Amoco's Solarex Venture to Merge with Enron Unit," by Timothy J. Mullaney, *Baltimore Sun* (December 20, 1994); "Solar Power, for Earthly Prices," *New York Times* (November 15, 1994; includes first Robert Kelly solar quotation); "Not Just a Marriage of Marketing Convenience . . . ," *Windpower Monthly* (June 1, 1997; for second Robert Kelly quotation and the quotation, "For some, the Enron-Zond deal was proof"); "Wind Reaps Fruit of Green Politics," *Windpower Monthly* (February 1, 1997). California wind farms' blade problems are described in "Windfarming in America," by Mike Evans. Sawin, "The Role of Government" (p. 605), gives the 2.4 percent and 13.5 percent figures on Enron/Zond manufacturing. TREIA's fall 1996 newsletter, p. 10, is the source of Kenneth Lay's quotation, "Renewable energy will capture." See also "Zond Machines to Fly in Texas' Delaware Mountain Wind Farm," *TREIA Newsletter* (Winter 1998–1999): 9.

Ward Marshall provided much of the recollection for CSW and the Fort Davis project; http://www.windpoweringamerica.gov/filter_detail.asp?itemid=695&print was also helpful.

7. A WIND REQUIREMENT

Information on the early days of the Wylys comes from interviews with Sam and Christiana Wyly, as well as from Sam Wyly's memoir, *1,000 Dollars and an Idea*. In 2006 Forbes calculated his worth at $1.1 billion, at forbes.com/lists/2006/54/biz _06rich400_samuel-wyly_G871.html.

The description of Bush's background is built on his *Decision Points*. Texas environmentalists' quotations on Bush's years as governor come from "Bush's Environmental Record," a PBS segment from August 22, 2000, at http://www.pbs.org/newshour/bb /election/july-dec00/bush_environment_8-22.html. The Wylys' campaign contribution information later in this chapter comes from a March 4, 2000, *Washington Post* story, "Texan Aired 'Clean Air' Ads; Bush's Campaign Not Involved, Billionaire Says," by John Mintz, p. A6.

Bush did not agree to an interview for this book, so his participation is reconstructed from numerous interviews and documents.

In an interview Pat Wood recalled his own childhood, his rise to the Public Utility Commission and beyond, and that crucial conversation with Bush. *Austin American-Statesman* reporter Bruce Hight's meticulous notes supplied us with the Wood quotation "I'm always the first to admit." Wood's wonderful self-description as a "young Reaganaut" comes from his remarks to the Natural Gas roundtable, Washington, D.C., May 26, 2005, at http://www.ferc.gov/eventcalendar/Files/20050527083106-05-26-05-pw .pdf. The deeper history of deregulation, especially during the Reagan administration, can be found in numerous books.

Wood's "I was not Mr. Tree Hugger" quotation comes from the *Austin American-Statesman* (June 3, 2001), "This Texan Has the Energy to Tackle Nation's Problems," by Bruce Hight.

TREIA newsletters, including spring 1996 ("Bush to Agency Heads—Pay Attention to Renewable Energy," p. 5), were helpful.

Our description of deliberative polling comes from interviews with James Fishkin, Ron Lehr, Karl Rábago, and Dennis Thomas; Fishkin, *When the People Speak*; "The Right Resource Mix," by Bruce Hight, the *Austin American-Statesman*, May 26, 1996; and "Texas Utility Deliberative Polls," an article by Ronald L. Lehr, Will Guild, and Dennis Thomas, in "Draft Report to the National Renewable Energy Laboratory," May 8, 2002 (http://www.repartners.org/tools/doc/TXUtilityDelPolls.doc). The TREIA newsletters also have information on deliberative polling in Texas. Other background information comes from "The Frontier Spirit: Democracy in Texas," in the May 16, 1998, edition of *The Economist*, and "Educated Opinions," by Larry Jones, *Electric Perspectives* 22, no. 1 (January/February 1997): 10.

A January 1997 Perryman Report on deregulation, published by the Perryman Group of Waco, gives a useful overview of efforts to regulate and deregulate Texas utilities. Statistics on the eighteen states deregulating as well as Texas's 1998 electric rate information come from "Power Switch: Texas Braces for Another Electricity Deregulation," an expansive article by Bruce Hight in the October 28, 1998, issue of the *Austin American-Statesman* (p. J1). The August 27, 1995, issue of the *Statesman* ("Bringing Competition to the State's Utilities; Chairman Pat Wood Will Oversee Fundamental Changes at PUC," by Bruce Hight, p. G1) contains the $15.5 million figure earned by public utilities in 1995. We also quoted from "PUC Commissioner Pat Wood II Sees Competition for Utilities," PUC newsletter, March 2, 1995; and a June 21, 1995, PUC memorandum from Wood to Commissioner Robert W. Gee. Bruce Hight's archives of the press releases from Wolens and Sibley were invaluable. We also quote from Hight's March 18, 1999, *Austin American-Statesman* story, "Texas Senate Passes Electric Deregulation Bill" (p. A1).

Wolens is identified as an "intellectual gladiator" in the title of a *Texas Monthly* story, "Intellectual Gladiator—Steve Wolens," from July 1999. Some David Sibley biographical information, including about his accident, comes from *Courage and Confidence* (New York: Simon & Schuster, 2010), by Karl Rove (p. 73).

A brief history of the biennial nature of the Texas Legislature is given in "Defying National Trend, Texas Clings to Biennial Legislature," by Kate Galbraith, *Texas Tribune* (December 31, 2010).

Further information on the haggling over Senate Bill 7 comes from interviews with Julie Blunden, Sam Enfield (who supplied the detail about the turbine being hauled in front of the capitol), Bob King, Jim Marston, Todd Olsen, David Sibley, Ed Small, Russel Smith, Tom "Smitty" Smith, V. John White, Steve Wolens, Pat Wood, and Sam Wyly.

Correspondence from Ken Lay and Sam Wyly to Bush comes from the George W. Bush papers at the Texas State Archives (including document 0294, the Wyly e-mail), as do the robo notes (box 2002/151–420: Correspondence—ROBOS—Energy—Renewable energy; from 2-17-99) and the notes from John Howard, a Bush advisor (box 2002/151–374: Policy Office correspondence 1999, Environment & natural resources, Howard, Western Governors Association and Wind Energy; memo date is 4/14/99, letter from Larry Jones to John Howard). Lay's testimony, found in Bruce Hight's notes, was before the House State Affairs Committee, April 9, 1998. "Enron's Gates Receives AWEA Award—Begins Association Presidency," *TREIA Newsletter* (Spring/Summer 1999): 12, provides Enron Wind's development information between July 1998 and June 1999.

Copies of the bills and their drafts are online at http://www.legis.state.tx.us/. TREIA newsletters recount the fate of the 2 percent renewables efforts in 1993.

Data on Texas's renewables and wind-energy capacity in 1999 is from several sources, including a Wolens release from Hight's collection. Data on the 1999 supply of wind on the U.S. grid comes from Sawin, "The Role of Government" (p. 19, citing a 2000 Renewable Energy Policy Project report by Marshall Goldbert), and "Global Wind Energy Market Report," 1999, found on AWEA's Web site at http://archive.awea .org/pubs/documents/globalmarket1999.html. The *Windpower Monthly* article cited (projecting wind's being 3 percent of Texas electricity by 2009) is titled "Texas Passes Law for Big Renewable Energy Portfolio" (July 1, 1999).

Ken Lay's letter to Bush citing Enron's status as "one of the largest wind turbine manufacturers in the world" is posted on Robert Bradley's Web site, http:// www.politicalcapitalism.org/enron/wind_PTC_2.jpg), and is also found in the Bush archives.

8. THE NEXT DECADE: TAKEOFF

Descriptions of Nolan County and Sweetwater come from A. C. Greene's *A Personal Country*, the invaluable *Handbook of Texas Online* (entries on Nolan County and Sweetwater), and the essay "In Windy West Texas, an Economic Boom," by Ben Block of the Worldwatch Institute, July 24, 2008, at http://www.worldchanging.com/archives/008271.html. We also cite Scarborough's 1925 novel, *The Wind*; "Winds of Change Blow in West Texas," CBS News, February 11, 2009; "The New Wildcatters," by Jennifer Bogo, published in PopularMechanics.com, December 2009, pp. 91–130, noninclusive (online at http://www.heliovolt.com/files/the-new-wildcatters.pdf); and "Texas Is More Hospitable Than Massachusetts to Wind Farms," by Kate Galbraith, *Boston Globe* (September 25, 2006; includes the Johnny Ussery anecdote).

Interviews with Andy Bowman, Chris Crow, Sam Enfield, Beth Garza, Mike Sloan, Randy Sowell, and Greg Wortham were helpful throughout this chapter.

Information on Trent Mesa and its opening comes from "Wind Farm Construction on Schedule in Trent," by Larry Zelisko, *Abilene Reporter-News* (June 5, 2001), and several press releases from the Trent Mesa Web site, http://www.trentmesa.com/news.htm, including "AEP and TXU Agree to Expand West Texas Wind Project" (July 18, 2001); "Trent Mesa Wind Project Dedication November 15" (November 15, 2001); "AEP Plans Strategic Growth in Renewable Generation; Recently Completed Wind Farm One of Nation's Largest" (November 20, 2001); and "GE Completes Enron Wind Acquisition; Launches GE Wind Energy" (May 10, 2002).

Information on Enron Wind as one of the six largest turbine manufacturers in the world in 2000 comes from Sawin, "The Role of Government" (p. 357). For Enron's collapse and its effect on the wind business, interviews with Ward Marshall and Tom Gray were helpful. Specific financial details on the Enron Wind sale come from "G.E. to Buy Enron Wind Assets," *New York Times* (April 12, 2002), as well as a June 27, 2001, e-mail from Martin Sosland, a lawyer involved in the deal.

McCamey information comes from the *Handbook of Texas Online* ("McCamey"), and interviews.

Additional Sweetwater information comes from "Nolan County: Case Study of Wind Energy Economic Impacts in Texas," by New Amsterdam Wind Source and West Texas Wind Energy Consortium, July 9, 2008 (http://cleanenergyfortexas.org/downloads/Nolan_County_case_study_070908.pdf); and "Wind Farm Money Fuels Spending in West Texas Schools," by Morgan Smith, *Texas Tribune* (November 11, 2011).

For more on Governor Perry, see "Gov. Rick Perry's Remarks Regarding Wind Energy—Amarillo," a press release dated October 27, 2006; and "Plugging in the Panhandle," by Kevin Welch, *Amarillo Globe-News* (October 29, 2006).

On transmission and the Hill Country battles, we drew from "In the Line of Ire," by Asher Price, *Austin American-Statesman* (July 22, 2009); *Handbook of Texas Online* ("Hill Country"); "Big Wind from Texas," by Kermit Pattison, *Fast Company* (July 14, 2008); "In Texas, Perry Has Presided Over Wind, Gas Booms," by Kate Galbraith, *Texas Tribune* (August 21, 2011; this also has Paul Sadler's quotation on Perry); "Perry Announces Major Energy Diversification Plan," press release from the Texas Governor's Office, October 2, 2006; "Controversial Hill Country Power Lines Canned," by Kate Galbraith, *Texas Tribune* (November 10, 2011); "Town Wages Artful Opposition to Power Line," by Asher Price, *Austin American-Statesman* (November 10, 2010); and

"Cost of Transmission Lines Nears $7 Billion," by Kate Galbraith, *Texas Tribune* (August 24, 2011).

9. THE FUTURE

The description of T. Boone Pickens in Pampa relies on "A Mighty Wind," by Asher Price. His Pampa lease update and natural gas comments come from "Boone Pickens on the Oil Spill, Gas and Wind Power," by Kate Galbraith, *Texas Tribune* (June 10, 2010). For Zilkha information, see "FAQ Regarding Goldman Sach's Acquisition of Zilkha," March 21, 2005, online in Google cache.

Descriptions of Herman Schellstede are drawn from "The Wind Over the Waves," by Erin Wayman, *GeoTimes* (April 2008); "A Few Snags, but Hopes Are Still High for Offshore Wind in Texas," by Kate Galbraith, *New York Times* (October 10, 2008); "Designer of Clean-up Ship Rues Missed Opportunity," by Harvey Morris, *Financial Times* (May 12, 2010); "Texas Wind Power Grows Along the Gulf Coast," by Kate Galbraith, *Texas Tribune* (February 11, 2011); and a 2012 interview with Herman Schellstede.

Jerry Patterson's comments come from "Blowing Strong," *The Economist* (June 29, 2006); "Big Wind from Texas," by Pattison. "Remoteness a Hurdle in Harvesting Panhandle's Winds," *Texas Tribune* (September 10, 2010), by Kate Galbraith, discusses the Panhandle, and an interview with Mike Sloan was helpful.

King Ranch background is found at "The Next Frontier," by S. C. Gwynne, *Texas Monthly* (August 2007), and Ward Marshall's response is from "Spread of Wind Farms Fans Fears That Regulation Isn't Keeping Up," by Asher Price, *Austin American-Statesman* (January 14, 2008). See "Dust-up Over Wind Hits Capitol," by Robert Elder, *Austin American-Statesman* (March 28, 2007): A1, for more on the King-Kenedy dispute.

10. THE LESSONS OF TEXAS WIND

The Nelson/Spinhirne tale comes from Naar, *The New Wind Power* (pp. 70–72). Nelson's quotation ("When the wind farm boom") is found in "Wizards of Wind," by Darcy Lively, *The West Texan* (Spring 2010): 8. Perry's quotation is from "Text of Governor Perry's Remarks at Dedication of Desert Sky Wind Farm," May 3, 2002, found on the Texas Governor's Office Web site. We interviewed Aron Brackeen and used material from "'Green Collar' Job Creation Is Conference's Goal," *Austin American-Statesman*, by Asher Price (February 17, 2009): B1. The jobs figure is found in "The Perryman Report & Texas Letter," June 2010, p. 2.

Andy Swift's comment is from "New Bachelor of Science in Wind Energy Announced," an August 5, 2011, press release from Texas Tech University. For Paul Sadler's salary, we looked at lobbying records at the Texas Ethics Commission, and we drew on Galbraith and Price, "A Mighty Wind." The Wortham quotation on solar is from "Summit Aims to Stir Up Solar Energy Talks," by Justin Zamudio, *San Angelo Standard-Times* (April 19, 2012). ERCOT provided us with figures for Texas grid generation, and the EIA has national figures. See "In Texas, Ambivalence Over Wind Tax Credit's Extension," by Kate Galbraith, *Texas Tribune*, February 14, 2012, for details on Texas officials' hesitation. Elder, "Dust-up Over Wind Hits Capitol," supplied the Dale Rankin anecdote.

Russell Gold's article is titled "Natural Gas Tilts at Windmills in Power Feud," *Wall Street Journal* (March 2, 2010). For the transport challenges of wind turbines, see "Slow, Costly and Often Dangerous Road to Wind Power," by Kate Galbraith, *New York Times* (July 22, 2009). We also cite "Texas Wind Energy Fails Again," by Robert Bryce, *National Review* (August 29, 2011), and we draw directly in this chapter from interviews with Tom Gray, Jim Marston, and V. John White.

POSTSCRIPT

Our information comes from interviews and e-mails with the sources involved, widely available news articles about the Wylys, and "Former President Bush Addresses Wind Conference," by Kate Galbraith, *Texas Tribune* (May 25, 2010).

BIBLIOGRAPHY

BOOKS

Asmus, Peter. *Reaping the Wind: How Mechanical Wizards, Visionaries, and Profiteers Helped Shape Our Energy Future.* Washington, D.C.: Island Press, 2001.

Baker, T. Lindsay. *A Field Guide to American Windmills.* Norman: University of Oklahoma Press, 1985.

———. *American Windmills: An Album of Historic Photographs.* Norman: University of Oklahoma Press, 2007.

———. *Blades in the Sky: Windmilling through the Eyes of B. H. "Tex" Burdick.* Lubbock: Texas Tech University Press, 1981.

Banks, Jimmy, and John E. Babcock. *Coralling the Colorado: The First Fifty Years of the Lower Colorado River Authority.* Austin, Tex.: Eakin Press, 1998.

Barta, Carolyn. *Bill Clements: Texian to His Toenails.* Austin, Tex.: Eakin Press, 1996.

Birkes, Darlene, Eloise Lane, and Elleta Nolte. *Gray County Heritage.* Dallas: Taylor Publishing Co, 1985.

Black, Edwin. *Internal Combustion: How Corporations and Governments Addicted the World to Oil and Derailed the Alternatives.* New York: St. Martin's Press, 2006.

Bomar, George W. *Texas Weather.* Austin: University of Texas Press, 1983.

Bush, George W. *Decision Points*. New York: Crown Publishers, 2010.

Caldwell, Gail. *A Strong West Wind*. New York: Random House, 2006.

Caro, Robert A. *The Years of Lyndon Johnson: Means of Ascent*. New York: Vintage Books, 1990.

———. *The Years of Lyndon Johnson: The Path to Power*. New York: Alfred A. Knopf, 1982.

Cray, Ed. *Ramblin' Man: The Life and Times of Woody Guthrie*. New York: W. W. Norton, 2004.

Egan, Timothy. *The Worst Hard Time: The Untold Story of Those Who Survived the Great American Dust Bowl*. Boston and New York: Mariner Books, 2006.

Fishkin, James S. *When the People Speak: Deliberative Democracy and Public Consultation*. Oxford: Oxford University Press, 2009.

Freeman, David S. *Energy: The New Era*. New York: Vintage Books, 1974.

Gipe, Paul. *Wind Energy Comes of Age*. New York: John Wiley and Sons, 1995.

———. *Wind Power: Renewable Energy for Home, Farm and Business*. White River Junction, Vt.: Chelsea Green Pub. Co., 1993.

Greene, A. C. *A Personal Country*. New York: Alfred A. Knopf, 1969.

Guthrie, Woody. *Bound for Glory: The Hard-Driving, Truth-Telling Autobiography of America's Great Poet-Folk Singer*. New York: E. P. Dutton, 1943.

Gwynne, S. C. *Empire of the Summer Moon: Quanah Parker and the Rise and Fall of the Comanches, the Most Powerful Indian Tribe in American History*. New York: Scribner, 2010.

Haley, J. Evetts. *The XIT Ranch of Texas and the Early Days of the Llano Estacado*. Norman: University of Oklahoma Press, 1929 (2nd printing March 1954).

Haragan, Donald R. *Blue Northers to Sea Breezes: Texas Weather and Climate*. Dallas: Hendrick-Long Publishing, 1983.

Harris, Coy F., ed. *Windmill Tales: Stories from the American Wind Power Center*. Lubbock: Texas Tech University Press, 2004.

Hills, Richard Leslie. *Power from Wind: A History of Windmill Technology*. Cambridge: Cambridge University Press, 1994.

Hirshberg, Gary. *The New Alchemy: Water Pumping Windmill Book*. Andover, Mass.: Brick House Publishing, 1982.

King, Robert J. *Alternatives to the Energy Crisis*. Governor's Energy Advisory Council Policy Analysis & Forecasting Division, 1977.

Koster, Rick. *Texas Music*. New York: St. Martin's Griffin, 2000.

Lewis Research Center. *Wind Energy Developments in the 20th Century*. Cleveland, Ohio: NASA, 1979.

McCarthy, Cormac. *All the Pretty Horses*. New York: Vintage Books, 1993.

McMurtry, Larry. *In a Narrow Grave: Essays on Texas*. Austin, Tex.: Encino Press, 1968.

———. *The Last Picture Show*. New York: Scribner Paperback Fiction, 1994.

———. *Leaving Cheyenne*. New York: Pocket Books, 1962 (reprint 1992).

Naar, Jon. *The New Wind Power: A Timely Firsthand Report on How Business, Government, and Independent Research Are Creating a New Energy Industry*. Foreword by Robert Rodale. Middlesex, Eng.: Penguin Books, 1982.

Nelson, Vaughn. *Wind Energy: Renewable Energy and the Environment*. Boca Raton, Fla.: CRC Press, 2009.

————, and Earl Gilmore. *Potential for Wind Generated Power in Texas*. Austin, Tex.: Governor's Energy Advisory Council, 1974.

Olien, Roger M., and Diana Davids Olien. *Oil in Texas: The Gusher Age, 1895–1945*. Austin: University of Texas Press, 2002.

————. *Wildcatters: Texas Independent Oilmen*. Austin: Texas Monthly Press, 1984.

Osborne, Michael J. *Beyond Light and Dark*. Austin, Tex.: Plain View Press, 2009.

Pierce, Ray. *Wind Energy: A High Plains Journal Book*. Dodge City, Kans.: High Plains Publishers, 1978.

Proulx, Annie. *That Old Ace in the Hole*. New York: Scribner, 2002.

Rathjen, Frederick William. *The Texas Panhandle Frontier*. Austin: University of Texas Press, 1973.

Raymond, Virginia Marie. "'The Wind' in the Literary Creation of West Texas." Master's thesis, University of Texas at Austin, 2003.

Righter, Robert W. *Wind Energy in America: A History*. Norman: University of Oklahoma Press, 1996.

————. *Windfall: Wind Energy in America Today*. Norman: University of Oklahoma Press, 2011.

Robinson, Roxana. *Georgia O'Keeffe: A Life*. Lebanon, N.H.: University Press of New England, 1989.

Rodenberger, Lou Halsell, Laura Payne Butler, and Jacqueline Kolosov, eds. *Writing on the Wind: An Anthology of West Texas Women Writers*. Lubbock: University of Texas Press, 2005.

Sawin, Janet Laughlin. "The Role of Government in the Development and Diffusion of Renewable Energy Technologies: Wind Power in the United States, California, Denmark and Germany, 1970–2000." PhD dissertation, Fletcher School of Law and Diplomacy, 2001.

Scarborough, Dorothy. *The Wind*. Austin: University of Texas Press, 1979 (orig. published 1925).

Spikes, Nellie Witt, with Geoff Cunfer, ed. *As a Farm Woman Thinks: Life and Land on the Texas High Plains, 1890–1960*. Lubbock: Texas Tech University Press, 2010.

Stegmaier, Mark J. *Texas, New Mexico, and the Compromise of 1850: Boundary Dispute and Sectional Crisis*. Kent, Ohio: Kent State University Press, 1996.

Virtus Energy Research Associates. *Texas Renewable Energy Resource Assessment*. Report for the Texas Sustainable Energy Development Council, 1995.

Webb, Walter Prescott. *The Great Plains*. New York: Grossett & Dunlap, 1931.

Wyly, Sam. *1,000 Dollars and an Idea*. New York: Newmarket Press, 2008.

Yergin, Daniel. *The Prize: The Epic Quest for Oil, Money & Power*. New York: Free Press, 1991.

ARTICLES

Particularly useful articles and other material

Burka, Paul. "Power Politics." *Texas Monthly*, May 1975.

Butts, David. "Renewed Effort for Renewable Energy." *The Texas Observer*, August 31, 1984, pp. 10–11.

Dingus, Anne. "The Windmill." *Texas Monthly*, January 1990, p. 112.

Eggleton, Claude W. "Marcellus L. Jacobs (1903–1985)." *Wind Power Digest*, no. 27, 1985, pp. 2–3.

Galbraith, Kate, and Asher Price. "A Mighty Wind." *Texas Monthly*, August 2011.

Gipe, Paul. "Lone Star Wind." *Wind Power Digest*, Fall 1979. (See multiple articles, pp. 36–48.)

Hight, Bruce. "West Texas Wind Power Plant Back Up to Full Speed." *Austin American-Statesman*, May 1, 1996, p. D4.

Jordan, Terry G. "Windmills in Texas." *Agricultural History*, vol. 37, no. 2 (April 1963), pp. 80–85. Accessed at http://www.jstor.org/stable/3740779, 1/16/2011.

Lindsley, E. F. "This Steam Car Works . . . and It Meets Tough '77 Emissions Standards." *Popular Science*, October 1974, pp. 74, 136–137.

"Something New under the Sun." *Dallas Morning News*, June 15, 1986, p. 30A.

"Wind Energy Conversion Systems: Workshop Proceedings." Washington, D.C., June 11–13, 1973. Accessed at http://books.google.com/ebooks/reader?printsec=frontcover&output=reader&retailer_id=android_market_live&id=dLcvAAAAYAAJ.

For more articles, see the Notes.

HELPFUL JOURNALS, NEWSLETTERS, AND FILM

Spectra: The Official Publication of the Texas Solar Energy Society. Elgin, Tex. Quarterly issues, 1982–1983.

Texas Renewable Energy Industries Association (TREIA) newsletters (1980s–1990s).

Windmillers Gazette, a publication overseen by T. Lindsay Baker.

Andrews, J. C. *Lease the Wind* (film, c. 1982).

Windpower Monthly

INTERVIEWS

Interviews form the heart of this book. We spoke with and/or e-mailed the following persons between 2009 and 2012: Jay Banner, Julie Blunden, Andy Bowman, Bob Bracher, Aron Brackeen, Robert Bradley, Mark Bruce, Jay Carter Jr., Jay Carter Sr., Matt Carter, Nolan Clark, Caroline Crimm, Chris Crow, Mike Crowell, Robert Cullick, Sam Enfield, John Fainter, James Fishkin, Tom Foreman, Beth Garza, Earl Gilmore, Carlos Gottfried, Tom Gray, Coy Harris, Bill Hobby Jr., Milton Holloway, Adam Holman, Walter Hornaday, Joe James, Bob King, Tony Kunitz, Ron Lehr, Ward Marshall, Jim Marston, Garry Mauro, Victor Murphy, Vaughn Nelson, Bill Neiman, Todd Olsen, Dale Osborn, Michael Osborne, Hal Phelps, T. Boone Pickens, Karl Rábago, Mark Rose, Herman Schellstede, Max Sherman, David Sibley, Michael Skelly, Mike Sloan, Ed Small, Russel Smith, Tom "Smitty" Smith, Martin Sosland, Randy Sowell, Kenneth Starcher, Dennis Thomas, Carol Tombari, Delbert Trew, Michael Webber, V. John White, Steve Wolens, Pat Wood, Chock Woodruff, Patrick Woodson, Greg Wortham, Christiana Wyly, Sam Wyly, and José Zayas.

INDEX

Note: Numbers in italics refer to illustrations.